Charles Codrington Pressick Hobkirk

A Synopsis of the British Mosses

Being Descriptions of all the Genera and Species Found in Great Britain and Ireland

to the Present Date. Second Edition

Charles Codrington Pressick Hobkirk

A Synopsis of the British Mosses
Being Descriptions of all the Genera and Species Found in Great Britain and Ireland to the Present Date. Second Edition

ISBN/EAN: 9783337058562

Printed in Europe, USA, Canada, Australia, Japan

Cover: Foto ©berggeist007 / pixelio.de

More available books at **www.hansebooks.com**

SYNOPSIS OF THE BRITISH MOSSES.

LONDON:
PRINTED BY GILBERT-AND RIVINGTON, LIMITED,
ST. JOHN'S SQUARE.

A

SYNOPSIS

OF THE

BRITISH MOSSES

BEING DESCRIPTIONS OF ALL THE GENERA AND
SPECIES FOUND IN GREAT BRITAIN AND
IRELAND TO THE PRESENT DATE.

BY

CHAS. P. HOBKIRK, F.L.S.

MEM. EFF. DE LA SOC. ROY. DE BOTANIQUE DE BELGIQUE; MEMBER OF THE
CRYPTOGAMIC SOCIETY OF SCOTLAND, AND OF THE MANCHESTER
CRYPTOGAMIC SOCIETY, ETC. ETC.

SECOND EDITION,

REVISED, CORRECTED, AND ENTIRELY REARRANGED.

LONDON:
L. REEVE AND CO.,
5, HENRIETTA STREET, COVENT GARDEN.
1884.

PREFACE TO FIRST EDITION.

It is not my desire that this little volume should be looked upon as anything more than what is expressed in the title, simply "A Synopsis of the British Mosses," and as a kind of *vade mecum* to the working Bryologist, as well as a guide to beginners. It is not altogether an original work, nor yet is it a mere compilation, for nearly every species has been carefully examined under the microscope before being described, and then the diagnoses compared with other works, principally that great text-book of British Bryologists, Wilson's "Bryologia Britannica." Besides this work, I have also largely consulted, and drawn from, Bruch and Schimper's "Bryologia Europæa," Schimper's "Synopsis," Dr. Mueller's "Synopsis," the Proceedings of the Linnean Society, the *Bulletins* of the Royal Botanical Societies of France and of Belgium; and last, but not least, the valuable papers recently contributed by Dr. Braithwaite to "Journal of Botany," "Grevillea," and the "Monthly Microscopical Journal," and also some papers by Mr. Mitten in the first-named publication.

In the general arrangement of the genera and species I have mainly followed the "Bryologia Britannica," as I did not consider myself justified in departing widely from it, although many of our principal Muscologists look upon it as very faulty; but I did not hold my authority sufficient to alter what has become a classical

arrangement amongst us; and more particularly as
both Dr. Braithwaite and the Rev. J. Fergusson are
engaged upon more critical examinations, prior to the
publication of new and more natural arrangements.
The Analysis of the Genera is principally founded upon
the same part from Wilson, and is intended not as an
arrangement, but merely a key.

I must here express my gratitude and thanks to
those gentlemen who have so kindly assisted me in its
preparation, both with the loan or gift of specimens of
the rarer and newer species, and also for the diagnoses
received from several, where specimens were not at-
tainable.   Amongst these gentlemen I must specially
thank Dr. Hooker for his kind permission to use the
Herbarium specimens and Library at Kew, and Mr. J.
G. Baker, F.L.S., for his valuable assistance in doing
so; also Dr. Braithwaite, F.L.S.; Mr. J. Bagnall, of
Birmingham; Dr. F. Buchanan White, of Dunkeld;
Dr. Fraser, of Wolverhampton; Rev. J. Fergusson, of
New Pitsligo; Mr. Carruthers, F.L.S., of the British
Museum; Mr. G. E. Hunt, of Manchester; Mr. John
Sim, of Strachan; Mr. W. Galt, of Edinburgh; M. P.
Goulard, of Caen, Calvados; and, lastly, all those
gentlemen and ladies who so readily came forward as
subscribers to the number of upwards of 200, to assist
in the publication of the volume.

CHAS. P. HOBKIRK.

HUDDERSFIELD,
    February, 1873.

PREFACE TO SECOND EDITION.

In presenting this Second Edition of my Synopsis of British Mosses to the Bryological student, a few words are necessary in addition to the prefatory remarks already given.

In addition to the works previously referred to, I may mention that I have also consulted " De Notaris Briologia Italiana," the first seven parts of Dr. Braithwaite's " British Moss-Flora " (all that are yet issued), Dawson Turner's " Muscologia Hibernica," an occasional reference to Mitten's " Musci Indiæ Orientalis," and numerous papers, critical remarks, &c., in the " Revue Bryologique," and the " Naturalist."

It will be noticed that the arrangement and classification have been entirely revised and altered. This was a subject requiring much consideration and thought; but I finally decided to adopt the arrangement of families and genera proposed by Jaeger in his "Adumbratio Muscorum," as seeming to me the most natural and convenient, besides which, with a few alterations, it nearly coincides with that of the " London Catalogue of British Mosses," second edition, the first edition of which was compiled and arranged from the same source by my accomplished friend Mr. H. Boswell, of Oxford, and myself. The arrangement of the species under the genera has been done at my own discretion—whether well or ill remains to be seen—with one or two exceptions shortly to be named.

The question of giving localities, even for the rarer species, was one to which I have given much thought,

and I finally decided to make it quite a secondary consideration, and have, therefore, generally given only
indications, so as not to swell out the book to an inor-
dinate size, thereby increasing the price.

My warmest thanks are due and are hereby gratefully
tendered to Dr. Braithwaite, F.L.S., for his kind permission to make a full use of his " Monograph of the
Sphagnaceæ," and, so far as it is published, of his mag-
nificent work the "British Moss-Flora," which, for the
British Islands, will take rank alongside the "Bryologia
Europea," and I wish him health and strength to complete it. The genus Sphagnum is, with his consent, a
condensation of the former work, both as to arrangement and diagnoses, and I have gladly availed myself
of his permission with regard to the latter, particularly
as regards the Dicranaceæ, Andreæaceæ, &c. Almost
equal thanks must be given to my friend Mr. H. Boswell, for much valuable assistance on critical points,
for specimens kindly contributed and lent, and for
many useful suggestions; also to Mr. Jas. Bagnall, of
Birmingham.

I have also compared my diagnoses carefully with
the second edition of Schimper's "Synopsis" and
Mueller's "Synopsis," and where I have considered it
requisite, have added to or modified my previous descriptions on these lines, besides carefully examining
many of the species over again, so that every species
and genus has been very diligently revised and, I trust,
improved. With these few remarks therefore I
commit the book to British Bryologists, in the hope
that it may prove useful to them, and be a means of
directing others to the study of the British Mosses.

CHAS. P. HOBKIRK.

1st May, 1884.

CONTENTS.

br., *branches*.

br. l., *branch leaves*.

cal., calyp., *calyptra*.

caps., *capsule*.

fem., *female*.

fl., *flower*.

fr., *fruit or fructification*.

infl., *inflorescence*.

innov., *innovations*.

l., *leaves*.

m.m., *millimetres*.

ped., *pedicel* or *seta*.

per., perist., *peristome*.

per. l. and p. l., *perichœtial leaves*.

per. teeth, *teeth of peristome*.

perig. l., *perigonial leaves*.

st., *stem*.

st. l., *stem leaves*.

Bry. Brit., Bryologia Britannica (Wilson).

Bry. Eur., Bryologia Europea.

Br. M. Fl., British Moss-Flora, by Dr. Braithwaite.

Ed. 1, refers to first edition of this work.

Muel. Syn., Mueller's Synopsis.

Rev. Bry., Revue Bryologique.

Schp. Syn., Schimper's Synopsis, ed. 2.

ADDENDA ET CORRIGENDA.

Page 2, *for* Fam. 1. Weisseæ, *read* Weissiæ.

Page 12, *for* Fam. 3. Trichostomaceæ, *read* Trichostomeæ.

Page 23, *line* 4, *delete all after* " inner," *and instead read*—"16 filiform, carinate processes, as long as or longer than outer teeth."

Page 24, *line* 10 *from bottom, for* Tetradontium *read* Tetrodontium.

Page 47, *line* 10, *delete* comma *after* apex.

Page 57, *line* 9, *for* Wheeldale *read* Wheldale.

Page 71, *after line* 19 *add*—
   Var. δ. RUGIFOLIUM, *Boswell.* L. somewhat contorted or crisped,
and more strongly undulated—perhaps only a form.
   Stockton Forest, Yorkshire, 1842 (Spruce), again 1874 (Anderson).

Page 79, *after* 91. C. flexuosus, *add*—
   Var. β. PALUDOSUS, *Schp.* Taller and more robust. L. longer,
more distant, often purplish at base.
   Boggy heaths. Llyn Ogwen, 1874, and Loch Maree, 1875 (H.
Boswell); Barmouth, Cader Idris.

Page 152, *lines* 21, 22—*O. gracile* was found in Finisterre (France)
   by M. Tanguy, Jun., in 1880, the first time it has been found
   out of England.

The word "mamillate" has unfortunately been several times spelled
   with double m.

SYNOPSIS OF BRITISH MOSSES.

CONSPECTUS OF GENERA.

Division I. SACCOMITRIA.

Order I. *HOLOCARPÆ.*

Caps. bursting irregularly. Calyptra lacerate.

Tribe i. ARCHIDIACEÆ.

L. with a more or less elongate subula, nerved, smooth, laxly reticulated. Caps. globose, sessile. Spores large, smooth, many-sided.

Genus 1. **Archidium,** *Brid.* As above.

Order II. *SCHIZOCARPÆ.*

Caps. on a pseudopodium, dehiscent by valves.

Tribe ii. ANDREACEÆ.

Acrocarpous, perennial. Branches dichotomous when old. L. generally brown or blackish, never green, patent or secund, cells thick-walled, rotundate-hexagonal above, rectangular below, generally papillose. Calyptra mitriform, thin. Caps. dehiscing in 4—6 valves.

2. **Andreæa,** *Ehr.* Caps. dehiscent below the middle by perpendicular valves.

*fruit at tips of branch or stem.

B

Tribe iii.   SPHAGNACEÆ.

Erect. Branches in fascicles, partly patent, partly reflexed. St. without radicles, bark spongy of large hyaline cells. L. generally with pores and spiral fibres in the cells, those of the stem dissimilar to those of the branches. Male fl. amentaceous, female gemmiform. Caps. at first sessile, afterwards slightly exserted, globose, dehiscent by a spurious lid or valve. Spores four-sided.

3. **Sphagnum,** *Dill.* , As above.

DIVISION II.   **STEGOMITRIA.**

Section 1.   *ACROCARPI.*

Fruit terminal on the stem.

Tribe iv.   WEISSIACEÆ.

Perennial, creeping or erect, with innovations below the floral apex. Branches dichotomous or fastigiate. L. parenchymatose, basal cells hyaline, upper densely chlorophyllose, often papillose. Caps. pedicellate, erect. Perist. simple, rarely absent. Calyptra cucullate, rostrate. Spores generally minute.

Fam. 1.   **Weisseæ.**

L. narrow, linear, lanceolate, and subulate, nerve terete. Caps. varying from ovate to cylindrical, regular or subincurved, gymnostomous or perist. single, 16-toothed, teeth articulate.

4. **Systegium,** *Schp.* Pl. minute. L. narrow, opaque, minute, papillose, crisped, cells in upper part minutely quadrate, at base lax hyaline, nerve excurrent or vanishing in apex. Caps. immersed or on a very

short seta, erect, symmetrical, lid persistent. Perist. absent. Monoicous.

5. **Hymenostomum,** *R. Br.* L. comal much longer than the rest, all crisped or cirrhate when dry, opaque and minutely papillose, cells hexagonal-rotund, chlorophyllose, rectangular and empty at base. Caps. with mouth closed by a thin membrane, annulus simple, lid persistent with a long beak. Spores large, globose, rough. Monoicous.

6. **Gymnostomum,** *Bry. Eur.* St. erect. Br. dichotomous or fastigiate. L. small, larger above, lanceolate or linear-lanceolate, margin plane, concave, nerve prominent behind, excurrent or vanishing, cells minute, quadrate, elongate hexagonal at base. Caps. erect, ovate, subglobose or elliptical, symmetric, mouth open, without either teeth or membrane, lid deciduous, subulirostrate. Calyptra cucullate. Spores minute, smooth. Monoicous or dioicous.

7. **Anœctangium,** *B. and S.* Densely cæspitose, dichotomous. L. linear-lanceolate, larger above, nerved, opaque, densely papillose, cells minute, rotundate-hexagonal above, rectangular hyaline at base. Caps. oval, ovate, or subglobose, with a reticulated membrane at mouth. Perist. absent; lid oblique, rostrate. Calyptra cucullate, oblique, with a long subulate beak. Spores minute, smooth, ferruginous. Dioicous.

8. **Eucladium,** *B. and S.* Innovations dichotomous. L. narrowly lanceolate, rigid, thickly nerved, smooth, cells irregularly quadrate, ovate above, lax and hyaline at base. Caps. erect, ovate or oval on a longer seta, lid subulirostrate. Perist. t. oblique, confluent, linear-lanceolate, bi- or trifid.

9. **Gyroweissia,** *Schp.* Pl. low, slender. L. linear-

obtuse, spreading, scarcely crisped when dry. Caps. elongate-elliptical, lid with a short beak, annulus broad. Perist. absent or rudimentary. Dioicous.

10. **Weissia**, *Hedw.* Small, cæspitose, dichotomous. L. lanceolate or linear-lanceolate, subulate, crisped when dry, minutely papillose, nerved, apical cells minute, quadrate, chlorophyllose, basal rectangular, hyaline; per. l. scarcely sheathing. Perist. t. 16, more or less irregular, punctulate. Annulus none or very narrow. Monoicous.

11. **Dicranoweissia**, *Lindb.* Pl. taller, fastigiate. L. patent, flexuose, crisped when dry, smooth or very slightly papillose, apical cells quadrate, basal rectangular, broader at angles; per. l. sheathing. Caps. elliptic elongate, lid subulirostrate. Perist. t. perfect, lanceolate, bifid at apex.

12. **Rhabdoweissia**, *B. and S.* Pl. small, dichotomous, monoicous. L. long, narrow, crisped when dry, shortly papillose above, margin serrulate or erose, cells quadrate or hexagonal, basal hyaline, lax. Caps. 8-striate, 8-sulcate when dry, truncate and wide-mouthed when ripe, teeth from a broad base shortly subulate, verruculose, lid subulirostrate. Calyptra large, cucullate. Spores large, punctulate.

Tribe v.   DICRANACEÆ.

L. between broadly and narrowly lanceolate and lanceolate-subulate, smooth or papillose at apex, upper cells quadrate, lower hexagono-rectangular or linear, quadrate at angles, nerve broad. Calyptra large, cucullate. Caps. on a long seta, cernuous, rarely erect, regular or incurved and subarcuate. Perist. t.

16, lanceolate, transversely trabeculate, generally cleft, reddish, hygroscopic, vertically striolate at back.

Fam. 1. Pseudo-Dicranæ.

L. papillose and chlorophyllose above, opaque, cells minutely quadrate, scarcely dilated at base.

13. **Cynodontium,** *B. and S.* Pulvinate. L. flexuoso-spreading, crisped when dry, opaque, papillose towards apex, upper cells minute, quadrate, lower hexagono-rectangular. Caps. subincurved, with a tapering symmetric or strumose neck, sometimes striate, and when dry sulcate. Calyptra inflated cucullate. Perist. t. lanceolate, unequally cleft, always vertically striolate.

14. **Dichodontium,** *Schp.* L. from a sheathing base, squarrose, curved, but scarcely crisped when dry, opaque, strongly papillose, cells at middle base rectangular, quadrate towards margin, smaller quadrate at apex, densely chlorophyllose. Caps. solid, generally without neck. Perist. t. large, bi- or trifid to below middle. Dioicous.

Fam. 2. Dicranæ-veræ.

L. smooth, apical cells oblong, basal rectangular or elongate, often quadrate at angles.

15. **Trematodon,** *Mich.* Pl. low, cæspitose. L. lanceolate and subulate, smooth, laxly reticulate, nerved. Caps. incurved, elliptic or oblong, with a long generally strumose neck, lid subulirostrate. Calyptra long-beaked, cucullate, inflated. Annulus of 1 or 2 rows of cells. Perist. t. lanceolate, simple or bifid nearly to base, striolate and papillose. Spores large, rough. Monoicous.

16. **Dicranella,** *Schp.* Pl. generally small. L.

smooth, cells oblong and elongate, thinly chlorophyllose, oblong hexagonal towards apex, lax, rectangular at base. Caps. generally cernuous, sometimes
striate. Perist. t. large, regular, solid, bifid, densely
articulate below, with filiform crura, minutely granulose. Spores medium smooth. Dioicous, rarely
monoicous.

17. **Dicranum,** *Hedw.* Pl. taller and more showy,
tomentose. L. generally patent or falcato-secund,
smooth, rarely papillose, shining or opaque, nerve
sometimes excurrent, elongate-lanceolate or lanceolate-
subulate, apical cells lineal-oblong, basal elongate,
generally very narrow, quadrate at angles, sometimes
inflated; per. l. more or less sheathing. Caps. erect
or cernuous, rarely striate, with a spurious neck,
sometimes strumose, lid generally subulirostrate.
Perist. t. bifurcate, interruptedly trabeculate. Spores
small. Monoicous or dioicous.

18. **Dicranodontium,** *B. and S.* Densely cæspitose,
radiculose. L. erecto-patent or falcato-secund, setaceous, nerved, shining, fragile, areolation lax, excavate,
reddish at angles. Caps. on an upright arcuate seta,
oblong or elongate-cylindric, without striæ, annulus
absent, lid convexo-conical, aciculate. Calyptra cucullate, base entire. Perist. t. 16, long, unequally bipartite to base, crura lanceolate-filiform. Dioicous.

19. **Campylopus,** *Brid.* L. broadly nerved, nerve
often sulcate at back, basal cells uniformly rectangular,
generally dilated excavate at angles. Caps. oval,
regular, generally striate on an arcuate seta, lid subulirostrate.. Calyptra cucullate, fringed at base. Annulus of 1, 2, 3 rows of cells. Perist. as in *Dicranum.*
Dioicous.

Tribe vi. LEUCOBRYACEÆ.

Pl. whitish, spongiose, soft when moist, fragile when dry. Leaf cells dimorphous, narrow and chlorophyllous, broad, empty, and porose.

20. **Leucobryum**, *Hampe*. L. erecto-patent and secund, concave, from an erect base, lanceolate. Caps. cernuous, incurved, ovate and oblong, striate, deeply sulcate when dry, neck shortly strumose, annulus absent, lid subulirostrate. Perist. as in *Dicranum*. Spores small, globose, reddish. Dioicous.

Tribe vii. FISSIDENTACEÆ.

Pl. small, slender. L. distichous, with a conduplicate dorsal wing, nerve excurrent, or ending in apex, cells minutely parenchymatous, very chlorophyllose.

21. **Fissidens**, *Hedw*. St. simple or slightly branched. Caps. symmetric or obliquely incurved, erect or cernuous, lid large, with a long or short beak. Perist. simple, of 16 teeth, lineal-lanceolate, unequally bifid, densely articulate at base, vertically striolate and punctulate, crura filiform, horizontally inflexed when dry. Annulus narrow. Spores small, smooth.

Tribe viii. SELIGERIACEÆ.

Pl. very small, with innovations from below the floral apex. L. narrow, nerved, smooth, apical cells minutely quadrate, basal rectangular. Caps. erect, symmetric, rarely gymnostomous. Perist. t. 16, more or less perfect.

Fam. 1. **Seligeriæ.**

Calyptra cucullate. Caps. subglobose, wide-mouthed, with a distinct neck.

22. **Seligeria**, *B. and S.* L. lower small, remote,

upper much larger, in a coma, subulate, nerve thin, margin entire. Caps. subspherical, on an erect or arcuate seta, turbinate when empty, neck tumid, lid large, convexo-conical, thin-beaked. Perist. absent (*Anodus*), or 16 free, linear-lanceolate teeth, articulate, without a central line, inflexed, when dry reflexed.

Fam. 2. Brachydontæ.

Pl. with same habit as *Seligeriæ*, but with larger chlorophyllose leaves. Calyptra mitriform, lobed, subulate. Caps. without neck. Perist. t. less solid, papillose.

23. **Campylostelium**, *B. and S.* Calyptra subulate-mitriform, erect, equally lobed. Caps. oblong, on an erect or arcuate seta. Annulus broad. Perist. t. 16, long, unequally bifurcate, subulate, arcuate-incurved when dry. Spores minute.

24. **Brachydontium**, *Bruch.* Calyptra split on one side nearly to apex, oblique. Caps. on an erect seta, oblong, obsoletely striate, sulcate when dry, lid convex, with a straight subulate beak, margin crenulate. Perist. t. confluent at base, broad, short, truncate or apiculate, hyaline, remotely papillose. Spores small, pale. Monoicous.

Fam. 3. Blindiæ.

L. shining, cells narrow, rectangular and dilated at basal angles. Caps. immersed or exserted, subsphærical, with a tumid neck. Perist. absent, or of 16 teeth. Lid large, acutely rostrate.

25. **Blindia**, *B. and S.* Caps. immersed, gymnostomous, or exserted, with a perist. of 16 equidistant lanceolate entire teeth, apex generally bifid, smooth,

somewhat hygroscopic. Calyptra large; or small (*Stylostegium*).

Tribe ix. LEPTOTRICHACEÆ.

Pl. varying in size. L. subulate, from a lanceolate base, smooth, shining, upper cells narrowly rectangular or minutely quadrate, basal laxer, hexagono-rectangular. Caps. erect, symmetric, oval, or cylindric. Perist. t. long, bifid, with filiform crura, papillose, erect or slightly oblique.

Fam. 1.　**Bruchiaceæ.**

Annual or perennial, simple or branched. L. lanceolate-subulate, uppermost in a comal tuft, nerved, shining. Calyptra small, cucullate. Caps. ovate, more or less acuminate. Spores small.

26. **Pleuridium,** *Brid.* L. from an oblong base, lanceolate and lanceolate-subulate, apex remotely and obtusely serrate. Caps. ovate-globose or ovate, shortly apiculate, smooth, shining.

Fam. 2.　**Ditrichaceæ.**

27. **Ditrichum,** *Timm.* Pl. small, cæspitose, or tall and slender. L. lanceolate-subulate, smooth, glossy, cells narrowly rectangular, basal lax and hexagono-rectangular. Caps. oval or cylindric, generally erect on a slender seta. Perist. t. 16, longish, cleft quite to base, filiform, articulate, papillose. Spores very small, smooth.

Fam. 3.　**Ceratodontæ.**

L. linear-lanceolate or subulate, obsoletely or distinctly papillose or smooth. Perist. t. long, densely articulate at base, divided almost to base.

28. **Ceratodon,** *Brid.* Many times dichotomously

branched, fastigiate. L. lanceolate, flexuose when dry, opaque, upper cells minutely quadrate-hexagonal, laxer at base, more or less papillose above, nerve strong. Caps. ovate-oblong, elongate, striate, deeply sulcate when dry, slightly cernuous. Perist. t. solid, 16, deeply bifid, papillose, with the basal membrane extending above mouth of caps. behind the teeth.

Fam. 4. **Distichiaceæ.**

L. distichous, compressed, subulate, from a sheathing base, smooth, cells minutely quadrate above, lax hexagono-rectangular below. Antheridia long, subcylindric, archegonia, long-styled. Caps. erect or cernuous. Perist. t. irregular, narrow, plane.

29. **Distichium,** *B. and S.* Densely cæspitose, silky. St. l. exactly distichous, spreading subulate, from a semi-amplexicaul base, nerve broad. Caps. erect or cernuous, coriaceous, shining, lid conical. Perist. t. linear-lanceolate, sub-entire, or variously bifid and lacerate, punctured, hygroscopic. Spores small. Monoicous.

Tribe x. **POTTIACEÆ.**

Perennial, rarely annual, densely gregarious or cæspitose. Cell formation of l. parenchymatose, hexagono-quadrate, and rotundate, above more or less papillose or smooth, densely chlorophyllose, at base laxer, dilatate-rectangular or hexagono-rectangular, hyaline or chlorophyllose. Fl. gemmaceous. Calyptra cucullate, rarely mitriform or lobed, smooth. Caps. generally erect, symmetric, or slightly curved, varying from subglobose to cylindrical. Perist. rarely absent, simple, lineal-lanceolate, entire or bifid, with filiform crura.

Fam. 1. Phascæ.

St. simple or divided, rarely with innovations. L. generally papillose at back. Monoicous. Antheridia enclosed or naked in axils of upper leaves. Caps. globose, ovate, ovate-acuminate or obliquely rostrate, either with or without a columella, immersed or exserted. Calyptra cucullate or many-lobed.

a. Columella absent.

30. **Ephemerella,** *Muell.* L. densely chlorophyllose, cells dense, rectangular above, elongate, rhomboid, hyaline at base. Caps. subspherical, immersed, vaginula cylindric. Spores reniform, brown, granulose.

b. Columella present.

31. **Microbryum,** *Schp.* Minute, cæspitose, gemmiform. L. dense celled, nerve solid, minutely papillose on back at apex. Antheridia naked, axillar. Caps. ovate, erect, slightly exserted. Calyptra large, erect, many-lobed, deeply slit on one side.

32. **Sphærangium,** *Schp.* Minute, gemmiform, gregarious. L. scariose, very concave, ovate, or obovate, nerved, minutely papillose on both sides at apex, cells large, only slightly chlorophyllose. Antheridia enclosed in minute gemmiform perigonial leaves. Caps. more or less immersed, erect, or subpendulous, spherical. Calyptra erect, small, mitriform. Columella thick. Spores largish, round, finely granulose.

33. **Phascum,** *Linn.* Pl. more robust. St. simple, bi- or tripartite. L. broad or ovate-lanceolate, solid, nerved, apical cells dense, quadrate-hexagonal, basal hexagono-rectangular, hyaline, minutely papillose. Antheridia enclosed. Caps. more or less exserted, subglobose, ovate, or ovate-oblong, with an obtuse

beak. Calyptra cucullate. Columella persistent.
Spores small.

Fam. 2. Pottiæ.

L. smooth or more or less papillose, nerve terete.
Calyptra cucullate, rarely many-lobed, oblique. Caps.
on a longer or shorter seta, erect. Perist. absent, or
of 16 fragmentary or linear teeth, more or less perfect.

34. **Pottia,** *Ehr.* Annual or biennial. St. simple
or slightly divided, radiculose below. L. broadish,
obovate-oblong, acuminate, opaque, smooth, or slightly
papillose, basal cells lax, hyaline, nerve smooth or
rarely lamellate. Caps. erect. Perist. absent, or if
present, of 16 plane teeth, generally imperfect; lid
rostrate or muticous.

35. **Didymodon,** *B. and S.* Perennial. St. with
innovations below apex, sometimes dichotomous, radiculose at base of innovations. L. elongate and linear-
lanceolate, sheathing at base, apex more or less serrate,
cells lax, hyaline at base, smaller at apex, and densely
papillose. Caps. cylindric, symmetric, or arcuate, on
a long seta. Perist. t. plane, linear-lanceolate, confluent
at base, more or less cleft above, punctulate, rarely
papillose. Spores minute.

Fam. 3. Trichostomaceæ.

Cæspitose or pulvinate. L. larger above, papillose,
apical cells minute, chlorophyllose, basal lax, diaphanous. Perist. t. united at base in a membrane,
more or less filiform above, papillose, crura equal or one
fragmentary, upright, oblique, or twisted. Spores
large and rough, or small and smooth.

36. **Trichostomum,** *B. and S.* L. linear-lanceolate,
opaque, minutely papillose. Perist. t. with a narrow

basal membrane, cleft to base into two equal filiform crura, often fragmentary, sometimes slighty oblique but not twisted, erect when dry. Annulus rarely absent. Spores as above.

37. **Barbula,** *Hedw.* Perennial. Habit, growth, and leaves similar to last. Perist. of 16 teeth, divided into 32 long filiform crura, spirally twisted and contorted, inserted into a tesselated membranous tube at base.

Tribe xi. CALYMPERACEÆ.

Perennial, dichotomous, with innovations below floral apex, densely radiculose. L. lingulate, apical cells minute, rotundate, chlorophyllose, densely papillose, basal lax, smooth. Caps. erect, oblong, or cylindrical, lid with a long styliform beak. Perist. simple or double, rarely absent. Calyptra very large.

38. **Encalypta,** *Schreb.* Cæspitose. L. linear, lingulate, or spathulate, upper cells hexagonal, hyaline mixed with chlorophyllose, lower hexagono-rectangular. Caps. erect, oblong, or cylindric, regular. Calyptra large, smooth, extinguisher-shaped, completely covering the capsule, with a long upright beak. Lid with a similar beak.

Tribe xii. GRIMMIACEÆ.

Generally acrocarpous, very rarely cladocarpous, with innovations below the floral apex. L. opaque, lower cells hexagonal or linear, and sometimes sinuous, upper hexagono-rotundate or punctiform, densely chlorophyllose. Fl. gemmiform. Caps. erect, generally symmetric, on an upright or arcuate seta. Calyptra mitriform, large, rarely cucullate, smooth, or sulcate, frequently pilose. Perist. seldom deficient, simple or double, both teeth and cilia plane, rarely filiform.

Fam. 1. Grimmiæ.

Perist. single, of 16 teeth, transversely articulate, lanceolate, entire, or perforate, or divided to base in two filiform crura. Calyptra small, often rough or hairy.

39. **Grimmia,** *Ehr.* Pulvinate, rarely cæspitose, branches dichotomous, fastigiate. Calyptra sometimes small many-lobed, or large sub-mitriform, cleft at base, or lobed and simply cucullate. Perist. t. lanceolate, entire, and perforate, or bi- or trifid, papillose, very rarely short and truncate.

40. **Racomitrium,** *Brid.* Pl. generally taller, broadly cæspitose. L. elongate-lanceolate, muticous, or hair-pointed, concave, margin recurved, cells at apex narrow, minutely quadrate, or linear-sinuous, lower cells elongate-sinuous, narrow. Caps. oblong, cylindric, on an upright rarely curved seta, lid small, beak subulate. Calyptra conico-mitriform, lobed, often darker coloured and rough at apex. Porist. t. with a narrow basal membrane, more or less deeply cleft into two subequal filiform crura.

Fam. 2. Ptychomitriæ.

Calyptra large, mitriform, plicato-sulcate, naked. Perist. t. 16, lanceolate, entire, or perforate, or linear, divided into two filiform crura.

41. **Glyphomitrium,** *Schw.* Pl. small, pulvinate, scarcely branched. L. solid, ovate- and elongate-lanceolate, curved when dry, apical cells roundish, minute, basal rectangular. Caps. erect, subglobose, lid acutely conical. Annulus absent. Calyptra cleft

on one side, and lacerate or incised at base. Perist. t. lanceolate, approaching in pairs, entire, reflexed when dry. Spores large, smooth.

42. **Ptychomitrium,** *B. and S.* Pl. somewhat larger, fasciculato-cæspitose. L. long, crisped, nerved, apical cells quadrate or rounded, basal linear-elliptical or hexagonal, opaque. Caps. erect, symmetric, on a longer seta. Calyptra naked or scaly, deeply lobed, beak aciculate. Annulus broad. Perist. t. long, linear-lanceolate, cleft to base into two subulate or filiform crura, erect when dry, papillose. Spores small, smooth, or punctulate.

Fam. 3. **Zygodontæ.**

More or less pulvinate, radiculose, dichotomously branched. L. patent or squarrose, linear-oblong or lanceolate-cuneiform, apical cells minutely quadrate, densely chlorophyllose, papillose, basal lax, hexagono-rectangular, hyaline, nerved. Caps. immersed or ex-serted, oval-oblong, striate, with a long neck, lid with an oblique beak. Perist. absent, single or double, of 8 broad lanceolate teeth, bifid, alternate with filiform cilia when double. Calyptra cucullate, smooth.

43. **Amphoridium,** *Schp.* Pulvinate, yellow or brown-green above, blackish below. L. soft, carinate, crisped when dry. Calyptra small, fugacious. Caps. without teeth, on a short seta, urceolate when dry, deeply furrowed.

44. **Zygodon,** *Hook and Tayl.* L. squarrose or patent, incurved imbricate when dry, linear-oblong or lanceolate-spathulate, nerve vanishing in apex or shortly excurrent. Caps. oval-oblong, on a long seta, with a tumid neck, less distinctly striate. Perist absent, single or double.

Fam. 4.   **Orthotrichæ.**

Cæspitose or pulvinulate, branches dichotomous and fastigiate.   L. apical cells punctulate, densely chloro-phyllose, basal lax, narrow, hyaline.   Calyptra large, mitriform, plicate, often hairy, rarely inflated.   Caps. immersed or exserted, symmetric, generally striate. Perist. single or double, rarely absent; ext. t. 16, approaching in pairs; int. 8 or 16 cilia, or a torn membrane.

45. **Ulota,** *Mohr.*   L. flexuose, much crisped when dry, except *U. Hutchinsæ,* narrowly areolate at base. Calyptra deeply plicate, rough, with yellowish hairs, deeply incised at base.   Caps. striate, sulcate when dry, oval, exserted.

46. **Orthotrichum,** *Hedw.*   Pulvinate, rarely cæspitose, erect.   L. ovate or elongate-lanceolate, minutely pa-pillose above, rarely smooth, basal cells hexagono-rectangular, hyaline.   Calyptra campanulate-mitriform, incised at base, less plicate than the last, pilose or naked.   Caps. rarely exserted, immersed, 8—16, striate and sulcate when dry, or rarely smooth.

Tribe xiii.   SCHISTOSTEGACEÆ.

Pl. annual, growing from a persistent confervoid prothallus.   Sterile pl. frondiform, l. vertically in-serted, confluent at base, fertile pl. frondose or leafless below, and with a few horizontal small leaves above. Cells lax.   Caps. small, erect, subglobose, without perist., lid convex, columella thick.   Calyptra small, mitriform, or dimidiate.

47. **Schistostega,** *Mohr.*   The only genus.   As above.

Tribe xiv.   SPLACHNACEÆ.

Annual or perennial.   L. broadish, soft, with large cells, hexagono-rhomboid, only slightly chlorophyllose, nerved.   Caps. on a long seta, erect, regular, seated on an obconical or pyriform apophysis, lid convexo-conical.   Calyptra cucullate, mitrate, or dimidiate. Perist. 16 geminate or 8 bigeminate teeth, rarely absent.

Fam. 1.   **Tayloriæ.**

L. broad, succulent.   Calyptra conico-cylindric or mitriform, and cleft on one side.   Caps. subspherical, with a long neck, erect or slightly incurved, with or without peristome.

a. Perist. absent.

48. **Œdipodium,** *Schw.*   Pl. soft, succulent.   L. broadly obtuse, nerve thick, vanishing at apex.   Caps. erect, tapering below by a long neck into the thick seta.   Calyptra long, narrow, cucullate.

b. Perist. present.

49. **Dissodon,** *Grev.*   Pl. less succulent.   L. obtuse, obovate, spathulate.   Caps. erect or cernuous, with a shorter obconical neck.   Perist. of 16 geminate, linear-lanceolate, or truncate teeth.   Calyptra mitriform, cleft up one side.

50. **Tayloria,** *Hooker.*   Laxly cæspitose, dichotomous, radiculose.   L. erecto-patent, spathulate, acute, obtusely serrate at apex.   Caps. with a long tapering pyriform neck.   Perist. t. 16, entire or bifid, linear-lanceolate, connivent incurved, reflexed when dry. Calyptra as in last.

Fam. 2.   **Splachneæ.**

Calyptra minute, conico-cucullate, suberect or conical

c

erect. Caps. short, cylindric, on an obovate-conical or spherical apophysis.

51. **Tetraplodon,** *B. and S.* Compactly cæspitose, radiculose. L. elongate-lanceolate or ovate-oblong, with a subulate apex. Caps. on an obconical apophysis as long as itself. Perist. of 16 teeth, approximating in fours, afterwards in pairs, short, incurved, reflexed when dry.

52. **Splachnum,** *Linn.* Annual, loosely cæspitose, dichotomous, fastigiate. L. remote, spreading, broadly obovate-lanceolate, narrow towards base, entire or serrate, nerve scarcely excurrent. Caps. small, cylindric, on a larger obpyriform or globose differently coloured apophysis, lid convex. Perist. t. 16, in pairs, linear, reflexed when dry.

Tribe xv. DISCELIACEÆ.

Small, almost stemless. L. oblong-lanceolate, nerveless, cells lax, rhomboid, and hexagono-rectangular. Caps. subglobose, subcernuous, on a long seta. Calyptra cleft on one side from base to apex. Perist. 16 lanceolate, cleft, or perforate teeth.

53. **Discelium,** *Brid.* The only genus.

Tribe xvi. FUNARIACEÆ.

Annual or biennial. L. broad, with large, lax, hexagono-rhomboid cells, sparsely chlorophyllose. Caps. erect and regular or cernuous and gibbous. Calyptra dimidiate, lobed, or campanulate. Perist. absent or present.

Fam. 1. Ephemeræ.

Pl. minute. St. short, radiculose, from a green filamentose prothallus. Calyptra thin, campanulate, irregularly cut at base. Caps. subsessile or exserted, with or without columella. Perist. absent.

54. **Ephemerum,** *Hampe.* Prothallus persistent. L. ovate, oblong-lanceolate or lanceolate-ligulate, with lax hyaline cells, nerved or nerveless, margin serrate or inciso-ciliate. Caps. subsessile, columella absent.

55. **Physcomitrella,** *Schp.* Prothallus not persistent. Pl. minute. L. spreading, reflexed, thinly nerved, serrate. Caps. on a longer or shorter seta, globose, columella thick. Calyptra campanulate, fugaceous.

Fam. 2. **Funariæ.**

Pl. larger. L. obovate and spathulate-acuminate. Calyptra shortly dimidiate or 5-lobed. Caps. globose or ovate, with a tumid neck, with or without peristome.

56. **Physcomitrium,** *Brid.* Calyptra shorter than capsule, 5-lobed. Perist. absent, with or without annulus.

57. **Entosthodon,** *Schw.* Calyptra dimidiate, with a long beak. Caps. erect, rarely subcernuous, with the neck pyriform, lid small, convex. Annulus absent. Perist. either rudimentary or of 16 narrow articulate teeth, confluent at base.

58. **Funaria,** *Schreb.* Calyptra dimidiate. Caps. cernuous, turgid, pyriform with the neck, gibbous, lid convexo-conical. Perist. rudimentary or double, external t. linear-lanceolate, with the apices oblique and converging into a disc, internal of opposite lanceolate cilia.

Tribe xvii. BARTRAMIACEÆ.

L. narrow, lanceolate, or subulate, cells more or less dense or lax, rhomboid, quadrate, or minutely quadrato

above and lax hexagonal below, sometimes papillose.
Caps. cernuous or erect, oblong, with or without neck,
or more or less globose, with or without striæ. Perist.
rarely absent, single or double.

Fam. 1. Amblyodontæ.

L. soft, cells lax, hexagono-rhomboid. Caps. with a
long neck. Perist. double, ext. t. half the length of
the int. processes, incurved when dry.

59. **Amblyodon,** *P. Beauv.* L. lower remote, upper
clustered, widely lanceolate, denticulate at apex, cells
large, nerved nearly to apex. Caps. subpyriform,
cernuous, incurved, with a small oblique mouth. Int.
perist. free, of 16 narrow processes.

Fam. 2. Meesiæ.

L. cells minute, thickened, hexagono-quadrate, rarely
lax and thin. Caps. obovate or subglobose, cernuous,
lid small, oblique. Perist. double, 16 short obtuse
teeth, more or less united with inner, which is a membrane deeply divided into 16 narrow carinate processes
longer than the teeth, cilia absent or rudimentary.

60. **Meesia,** *Hedw.* Pl. tall, radiculose, and tomentose. L. lanceolate or subulate, upper in a terminal
cluster, obtuse, erect, or spreading, cells densely
hexagono-rectangular. Caps. obovate, with a long
neck tapering into the long seta. Perist. as above.

61. **Paludella,** *Ehr.* Pl. tall, strongly radiculose
and tomentose. L. squarroso-recurved, decurrent,
ovate-lanceolate, nerved nearly to apex, cells dense,
roundish, papillose. Caps. cernuous, oblong, on a long
seta, slightly curved, with a short neck. Perist. outer
teeth and inner processes of equal length.

62. **Catoscopium,** *Brid.* Pl. smaller. L. lanceolate, carinate, spreading, but only slightly recurved, nerved, cells minute, hexagono-quadrate. Caps. on a shorter seta, small, cernuous, subglobose. Perist. outer teeth short, obtuse, inner imperfect.

Fam. 3.   Bartramiæ.

L. narrowly lanceolate and subulate, apical cells minutely quadrate, basal lax, hexagonal, more or less papillose. Caps. globose, rarely ovate, generally striate and sulcate when dry, lid small. Perist. absent, simple or double, teeth lanceolate, internal processes bifid, on a narrow basal membrane, cilia inconspicuous. Male fl. discoid.

63. **Bartramia,** *Hedw.* St. erect, dichotomous. L. from a sheathing base, lanceolate or lanceolate-subulate, densely papillose, serrate at apex, nerve vanishing or excurrent, hispid on back at apex. Caps. on a straight or curved seta, erect or cernuous, globose or ovate-globose, striate, sulcate when dry, lid minute, convex, obtusely acuminate. Perist. absent or single, but generally double, outer 16 lanceolate teeth converging, inner a membrane of 16 plicæ divided above into 16 lanceolate carinate processes, sometimes with cilia. Annulus absent.

64. **Conostomum,** *Swartz.* St. erect, radiculose. Br. fastigiate. L. longer above, lanceolate and lanceolate-subulate, carinate, cell formation like *Bartramia.* Caps. oval, cernuous. Perist. single, of 16 equidistant linear-lanceolate teeth, united at apex into a cone.

65. **Bartramidula,** *B. and S.* Pl. small, slender, decumbent. Branches verticillate. L. erecto-patent, secund, lanceolate, denticulate, subpapillose, cells lax,

hexagono-rectangular, nerve vanishing. Caps. globose-pyriform, on a curved seta, without striæ. Perist. absent, lid small, plano-convex.

66. **Philonotis,** *Brid.* Pl. small and decumbent, or erect and larger. Br. at floral apex verticillate or fasciculate, the rest dichotomous, radiculose-tomentose. L. small, erect, sometimes secund, lanceolate, strongly serrate, cells as in *Bartramia.* Caps. on a long seta, cernuous, globose, striate, lid small, oblique. Perist. internal, with distinct binate cilia.

67. **Breutelia,** *Schp.* Pl. robust, irregularly branched, tomentose. L. somewhat sheathing, more or less broadly lanceolate, deeply 5-sulcate, nerve thin, vanishing. Caps. pendulous, on a curved seta, spherical or ovate-oblong, striate, sulcate when dry, lid small, mamillate.

Tribe xviii.   BRYACEÆ.

Perennial. L. cells either all parenchymatose or prosenchymatose above and parenchymatose below. Calyptra coriaceous, narrow, cucullate, mostly deciduous. Caps. on a long seta, with the neck long or shortly pyriform, or subspherical, erect, cernuous, inclined, or pendulous. Perist. single or double, mostly the latter, outer teeth trabeculate and lamellate within.

68. **Mielhichhoferia,** *N. and H.* L. lanceolate, smooth, serrate, cells uniformly hexagono-rhomboid or linear. Caps. erect or cernuous, pyriform or clavate, seta terminal on a branch. Perist. single, of 16 narrow slender teeth, confluent at base.

69. **Orthodontium,** *Schw.* Pl. small, cæspitose. L. narrow, flexuose, cells hexagono-rhomboid. Caps.

erect, with the neck oval or oblong. Calyptra cucullate. Perist. double, outer 16 lanceolate subulate teeth without medial line, inner a narrow membrane deeply divided into 16 short filiform carinate processes, cilia absent.

70. **Leptobryum**, *Schp.* Annual. L. narrow, upper ones longest, subulate, flexuose, cells upper linear-rhomboid, lower lax, hexagono-rectangular, pellucid, nerve broad. Calyptra small. Caps. inclined or pendulous, with a long neck, lid mamillate. Perist. as in *Bryum*.

71. **Webera**, *Hedw.* Pl. simple or branched, with innovations from the base. L. lanceolate, seldom broader, cells rhomboid-hexagonal or more or less linear, nerve terete. Caps. on a long seta, cernuous or horizontal, with a longer or shorter neck, obovate or pyriform. Annulus rarely absent. Perist. t. long, internal processes arising from a carinato-plicate membrane, with or without cilia, without appendages at the internodes.

72. **Zieria**, *Schp.* St. with innovations below the floral apex, radiculose to summit. L. imbricate, ovate, and oblong-acuminate, apiculate, or with nerve excurrent, margin entire, cells lax, hexagono-rhomboid. Caps. horizontal, or bent downwards and then upwards, on a sigmatoid seta, elliptic, lid small, convex, obliquely apiculate. Perist. t. narrowly lanceolate, int. membrane narrow, with narrow processes and rudimentary cilia.

73. **Bryum**, *Dill.* Perennial, with innovations from below the floral apex, generally radiculose. L. semi-amplicaul, subdecurrent, ovate- and oblong-lanceolate, acute, acuminate, rarely subrotund and obtuse, cells

hexagono-rhomboid, lax or denser.   Synoicous, monoicous, or dioicous.   Male fl. gemmiform, rarely discoid.
Calyptra narrowly cucullate.   Caps. inclined or pendulous, on a longer or shorter seta, including the neck
generally pyriform, regular or slightly curved, lid
convex.   Perist. double, ext. t. long, linear-lanceolate
or subulate, densely articulate below, strongly hygroscopic, int. membrane 16-carinate, half as high as the
teeth, with carinate processes, and cilia either fragmentary adherent to the teeth or free, 2- 3-nate,
filiform, with appendages at the internodes or articula-
tions.

Tribe xix.   GEORGIACEÆ.

Pl. cæspitose.   L. ovate or lanceolate, smooth, nerve
thin, cells hexagono-rotund, sparingly chlorophyllose.
Caps. erect, oval or cylindric.   Perist. of 4 pyramidal
teeth.   Annulus absent; lid conical.

74. **Tetraphis,** *Hedw.*   (*Georgia,* Ehr.) Cæspitose,
erect.   L. lower small, remote, upper larger, broadly
ovate-lanceolate, basal cells lax, nerve vanishing.
Calyptra large, covering half the caps., which is
cylindric on a long seta.

75. **Tetradontium,** *Schw.*   Pl. small, gregarious.   St.
short, with flagelliform innovations, and long, linear-
clavate, flattish processes.   L. few, ovate-lanceolate,
entire, almost nerveless.   Caps. erect, oval-oblong, on
a long rigid seta, lid conical.   Perist. as in *Tetraphis,*
but shorter.

Tribe xx.   MNIACEÆ.

Pl. tall, with innovations from the base or beneath
floral apex.   L. large, with large hexagono-rectangular
cells, often bordered and toothed.   Male fl. discoid

or gemmiform. Caps. inclined or pendulous, almost without neck, plane or striate. Perist. double, cilia of internal scarcely appendiculate.

Fam. 1. **Mniæ.**

Innovations from base of stem. Male fl. discoid.

76. **Mnium,** *Dill.* Gregarious or cæspitose. L. lower generally remote, small, much larger above, from a narrow sometimes decurrent base, obovate-spathulate or broadly oblong and lingulate, nerved, margin generally thickened, entire or singly or doubly toothed, rarely without a thickened margin. Calyptra cucullate. Caps. on a long seta, inclined or pendulous, ovate, oblong, or sub-globose, lid convexo-conical, mamillate. Perist. as in *Bryum,* but almost exappendiculate. Annulus present.

77. **Cinclidium,** *Swartz.* Habit of *Mnium.* Caps. pendulous, oval or oblong, with a short tumid neck. Calyptra minute, cleft on one side. Annulus absent. Perist. internal, a reticulated cupuliform membrane, carinato-plicate, perforated, divided above into 16 narrow carinate processes, and united at base to the outer teeth. L. large, roundish, bordered entire.

Fam. 2. **Aulacomniæ.**

Innovations from below floral apex. L. with apical cells, narrowly hexagono-rotund, laxer at base. Caps. cernuous, oblong, striate, sulcate when dry. Perist. as in *Mnium.*

78. **Aulacomnion,** *Schw.* L. loosely imbricate, oblong, elongate, and linear-lanceolate, papillose; male fl. discoid or gemmiform. St. frequently bearing *pseudopodia* of rudimentary leaves. Perist., internal processes with 2 or 3 long filiform cilia. Dioicous.

Fam. 3.   **Timmiæ.**

Innovations from below floral apex. L. elongate-
lanceolate from a sheathing base. Fl. monoicous,
gemmiform. Caps. horizontal or inclined, oval, oblong,
lid convex. Perist. double, internal a plane basal
membrane with numerous cilia.

79. **Timmia,** *Hedw.* L. all equal, elongate-lanceolate,
nerve strong, cells upper minutely hexagono-rotund,
basal lax, hexagono-rectangular. Caps. ovate or oblong, smooth or obsoletely striate. Perist. outer t.
when dry geniculato-incurved, cilia of inner nodoso-
filiform.

Tribe xxi.   BUXBAUMIACEÆ.

Pl. small, almost stemless, producing a large, oblique,
ventricose capsule, sessile or on a thick seta. Calyptra
small, conical. Perist. double, external more or less
perfect, internal a conical, twisted, plicate membrane.

80. **Buxbaumia,** *Haller.* Pl. small. L. ovate and
oblong-lanceolate at base of seta, margin grossly
dentate, nerveless, areolæ lax. Caps. very large, on a
rough, fleshy seta, oblique, convex on one side, flattish
on the other or upper, ventricose, lid conical, obtuse.
Calyptra very small, covering only the lid. Outer
perist. of 3—4 layers of cellular tissue, divided into
irregular or moniliform reddish teeth, inner a pale,
plicate, twisted, conical membrane.

81. **Diphyscium,** *W. and M.* St. short, simple. L.
lingulate, almost entire, spreading, flexuose, nerved,
margin plane, entire, or slightly denticulate at apex;
areolæ granular, opaque; per. l. longer, lanceolate, the
aristate apex divided or ciliate. ˏCaps. sessile, ovate-
conical, ventricose, oblique, gibbous below. Perist.

double, outer a filmy, narrow, whitish ring, quite rudimentary, internal an elongate, conical, 16-plicate membrane.

Tribe xxii.   POLYTRICHACEÆ.

Pl. firm, more or less rigid, varying in size. L. soliu, areolæ thick and dense, the nerve often forming the greater portion, lamellate on upper surface. Calyptra narrow, cucullate, spinulose at apex, rarely smooth, generally covered with long hairs. Caps. on an upright or cernuous seta, round or angular, more or less elongate. Perist. single, rarely absent, generally of very short lingulate teeth, rarely prolonged into cilia.

Fam. 1.   **Polytrichæ.**

Perist. of 16, 32, or 64 shortly lingulate teeth, united below by a membranous tympanum covering the mouth of the capsule.

82. **Oligotrichum**, *De Cand.*   St. simple, with sub-terraneous stolons. L. very concave, scarcely undulate, elongate and linear-lanceolate, margin incurved, almost tubulose at apex, remotely serrate, nerve terete and naked below, dilated and lamellate above. Caps. erect, ovate-oblong, rounded. Calyptra large, with a few upright hairs. Perist. of 32 short, rigid, obtuse teeth.

83. **Atrichum,** *P. Beauv.*   Pl. gregarious or cæspitose. L. lingulate, scarcely sheathing, more or less undulate, crisped when dry, margin acutely serrate, nerve narrow, slightly lamellate. Caps. oval or cylindric, cernuous, straight or slightly arcuate, lid hemispherical, with a long slender beak. Perist. as in last. Calyptra narrow, spinulose at apex only.

84. **Pogonatum,** *P. Beauv.*   St. short and simple, or

longer and dendroid. L. from a sheathing membranous base, patent, rarely subsecund, scarcely different when dry, elongate-lanceolate or lingulate, nerve very broad, occupying nearly the whole lamina, with many lamellæ, margin more or less sharply serrate. Caps. ovate-globose, not angular, erect or cernuous, when dry and empty spherical or cyathiform, lid convex, with an upright beak, sometimes concave when dry. Calyptra densely covered with hairs reaching below its base. Perist. as above.

85. **Polytrichum,** *Dill.* Pl. taller, erect. St. simple, ligneous, radiculose. L. lower squamiform, upper from a sheathing membranous base, horizontally spreading; nerve broad, dilated and occupying the width of the leaf above, covered with numerous lamellæ, sometimes excurrent into a bristly point; margin either incurved and entire, or plane and sharply serrate. Caps. on a long seta, upright, when dry horizontal, 4 or 6-angled, on a discoid or subglobose apophysis, lid large, convex or conical, with an upright or oblique beak. Perist. of 64, rarely 32 short teeth. Calyptra with long whitish or brownish hairs covering the whole capsule.

Section 2. CLADOCARPI.

Fruit terminal on a branch.

Tribe xxiii. FONTINALACEÆ.

Pl. floating. L. tristichous, or in 5—8 rows, cells prosenchymatous or parenchymatous. Fl. laterally inserted among the leaves. Caps. generally immersed, perist. double, ext. of 16 long linear teeth transversely articulate; int. cilia either quite free, or united into a

16-carinate cone, or single, of 32 teeth on a membranous base, interlaced below, free above. Calyptra mitriform or dimidiate.

Fam. 1. Cinclidotæ.

L. in 5—8 rows, parenchymatous, flexuose, nerve excurrent, areolæ small, opaque. Perist. single.

86. **Cinclidotus**, *P. Beauv.* L. patent, solid, nerve strong, excurrent. Caps. immersed on a short seta, ovate-oblong, sulcate when dry. Perist. large, t. more or less perfect, somewhat twisted.

Fam. 2. Fontinalidæ.

Calyptra conico-mitrate, vaginula imperfect. Caps. immersed in perichætium. Perist. double, internal a handsome clathrate 16-carinate cone. L. tristichous, cells prosenchymatous.

87. **Fontinalis**, *Dill.* St. slender. Br. fasciculate. L. deeply concave and carinate, the two blades being almost folded together, nerveless, cells narrowly rhomboid and vermicular, laxer in auriculate angles. Perist. double, as above. Annulus absent. Dioicous.

Fam. 3. Dichelymeæ.

L. narrow, thinly nerved. Caps. immersed or exserted. Perist. double, int. either a clathrate cone or of free cilia. Calyptra dimidiate or shortly cucullate, vaginula distinct.

88. **Dichelyma**, *Myr.* Pl. more or less distinctly pinnate-branched. L. long, narrow, generally subulate, subfalcate, cells rhomboid. Per. l. long, outer imbricate, inner convolute. Caps. and perist. as above.

Tribe xxiv. CRYPHÆACEÆ.

St. dichotomous or subpinnate. L. broad or ovate-acuminate, nerveless, cells at apex minute, at base larger, quadrate, or linear. Calyptra mitrate, cucullate, or conical, sometimes with apex roughened. Caps. immersed or exserted, gymnostomous, or perist. double.

Fam. 1. **Hedwigiæ.**

Br. dichotomous or irregular. L. broad, coriaceous, cells quadrate, minute at apex, larger at basal angles, and linear at mid-base. Calyptra mitrate or cucullate. Caps. gymnostomous.

89. **Hedwigia,** *Ehr.* Br. dichotomous. L. strongly papillose. Calyptra small, conico-mitrate. Caps. globose, immersed, lid broadly convex.

90. **Hedwigidium,** *B. and S.* St. irregularly branched, stoloniferous. L. solid, smooth. Calyptra bi- trilobed or cucullate. Caps. exserted, lid conical.

Fam. 2. **Cryphææ.**

Perich. branch often elongate. Calyptra conico-mitrate, roughened at apex. Caps. immersed, tumid at base. Perist. absent, single or double.

91. **Cryphæa,** *Mohr.* St. pinnate or rarely bipinnate, L. ovate, acuminate, upper cells minute, rhomboid. Calyptra small, conical, rough at apex. Perist. double, outer of 16 narrow converging teeth, inner 16 alternating, narrow, filiform, carinate processes.

Section 3. *PLEUROCARPI.*

Fruit on a lateral branch.

Tribe xxv. LEUCODONTACEÆ.

Secondary stems erect or pendulous. L. patent,

secund, or complanate. Calyptra cucullate, pilose, or smooth. Caps. shortly exserted or on a long seta, erect, regular. Perist. single or double.

Fam. 1. Leptodontæ.

L. complanate, broad, thinly nerved, cells minutely rhomboid. Vaginula distinct. Calyptra hairy. Perist. single, of 16 pale, slender teeth.

92. **Leptodon**, *Mohr.* Secondary stems short, depressed, simple or bipinnate. L. obtuse, bisulcate, opaque, smooth or slightly papillose, ovate-oblong or subrotund, scarcely nerved to apex. Caps. oval-oblong, lid with a straight beak. Perist. t. 16, slender, equidistant, minutely papillose, whitish. Annulus absent. Dioicous.

Fam. 2. Leucodontæ.

Secondary stems irregularly or subpinnatedly branched. L. patent, secund, lanceolate, sometimes sulcate. Per. l. much sheathing. Calyptra large, smooth, or scarcely hairy. Caps. on a longer or shorter seta, erect, regular. Perist. single or double, more or less perfect.

93. **Pterogonium**, *Sw.* Primary stem a creeping rhizome, secondary erect. Br. fasciculate or dendroid. L. spreading, imbricate when dry, smooth, broadly obovate, acuminate, serrulate, nerve bifid, cells narrow, obliquely oval at base, except in middle base, apical, shortly fusiform. Caps. regular or subarcuate. Calyptra very slightly hairy. Perist. t. large, densely articulate.

94. **Leucodon**, *Schw.* Secondary st. erect or arcuate, stolouiferous. L. decurrent, broadly ovate-lanceolate, deeply sulcate, nerveless, cells at apex and mid-leaf

linear-vermicular, at side base oblique, dot-like at
basal angles. Caps. ovate-oblong or subcylindric on
an upright seta. Perist. t. short, slender, bi- trifid,
remotely articulate, strongly papillose, pale or whitish.
Calyptra as large as capsule.

95. **Antitrichia,** *Brid.* Secondary st. decumbent and
pendulous. Br. subpinnate. L. nerved, sulcate, otherwise same as last. Caps. on a flexuose or upright seta.
Perist. double, outer t. narrowly lanceolate-subulate,
slender, entire, almost smooth, int. processes about
same length, subulate, slightly carinate, without basal
membrane. Calyptra shorter than caps.

Tribe xxvi.   NECKERACEÆ.

Secondary st. erect or pendulous, pinnate, rarely
bipinnate. L. broad, campanulate, scariose, generally
undulate, nerve very slender or absent, cells narrowly
linear-rhomboid. Calyptra cucullate or conical, cleft,
naked or hairy. Caps. immersed, rarely exserted,
erect, symmetric. Perist. single or double.

96. **Neckera,** *Hedw.* Calyptra naked or slightly
hairy, lid rostrate, annulus absent. Perist. double,
ext. t. long, lanceolate-subulate, int. processes short or
longer, filiform from a membranous base, cilia absent.

97. **Homalia,** *Brid.* L. distichous, complanate, unequal sided, thinly nerved. Vaginula elongate, hairy.
Calyptra cucullate, naked. Caps. cernuous on a long
seta, lid with a slender, oblique beak. Perist. double,
ext. t. lanceolate, linear-subulate, confluent at base,
int. processes carinate, from a broad carinato-plicate
basal membrane, with longer or shorter cilia.

98. **Thamnium,** *Schp.* Dendroid, from a creeping
rhizome. L. subcomplanate, nerved, apical cells small,

quadrate, and rhomboid, basal narrowly oblong. Caps. short, turgid, ovate, lid with a long beak. Perist. ext. t. long, densely lamellate, int. processes perforate, with long appendiculate cilia.

Tribe xxvii. HOOKERIACEÆ.

Pl. short or longer, branched. L. narrow, patent, secund or broad and complanate, singly or doubly nerved, cells lax. Per. l. scarcely sheathing. Calpytra conical or mitrate, almost entire at base, or lobed and fimbriate at base, papillose or squamous. Caps. on a smooth or papillose seta, cernuous or horizontal. Perist. large, double.

Fam. 1. Daltoniæ.

Pl. small, slender. L. patent, narrow, cells narrowly rhomboid, laxer at base. Calyptra fimbriate at base. Seta generally papillose. Perist. ext. t. linear-lanceo-late, subulate, int. processes narrowly subulate from a basal membrane.

99. **Daltonia.** L. ovate and linear-lanceolate, margin thickened, more or less flexuose when dry, singly nerved nearly to apex, lower cells oblong-hexagonal, hyaline, quadrate-dilated at base. Caps., calyptra, and perist., as above.

Fam. 2. Hookeriæ.

Pl. taller. L. narrow or broad, acuminate, obtuse, truncate, binerved, rarely nerveless, cells lax. Calyptra mitrate, lobed or entire. Caps. subhorizontal, lid beaked. Perist. ext. t. long, entire, solid, densely articulate, int. processes from a carinato-plicate basal membrane, carinate, and as long as outer teeth.

D

100. **Pterygophyllum,** *Brid.* L. broadly complanate, arge, obliquely and densely imbricate, smooth, ovate or acuminate, nerveless, margin scarcely bordered, entire. Caps. small. Calyptra mitrate, lobed. Perist. outer teeth reddish, with a single medial line, inner processes with rudiments of cilia.

101. **Hookeria,** *Tayl.* L. soft, nerve double, margin generally bordered, serrate. Calyptra shortly incised at base. Perist. outer t. with two reddish prominent ridges enclosing a yellowish furrow, int. processes entire, no trace of cilia.

Tribe xxviii.   FABBONIACEÆ.

Pl. small, creeping. L. patent or subsecund, ovate-lanceolate, nerveless or shortly nerved, smooth, cells rather lax, hexagono-rhomboid or rhomboid. Calyptra cucullate.   Caps. erect, regular or slightly curved. Perist. single or double, 8—16 teeth, internal a torn membrane or of filiform cilia.

102. **Habrodon,** *Schp.* L. nerveless, soft, squarrose, ovate-lanceolate, longly and acutely acuminate, entire ; per. l. eroso-denticulate, hyaline.   Caps. narrowly oblong, lid conical, muticous.   Perist. t. linear-lanceo-late, connivent when dry, 8 bi-geminate or 16 geminate, int. articulate, filiform cilia without basal membrane. Annulus narrow, hyaline.

103. **Myrinia,** *Schp.* Cæspitose, pulvinate. L. ovate-lanceolate, not equilateral, entire, obsoletely nerved, smooth. Caps. oblong, subcylindric, regular or slightly incurved, lid conical. Annulus absent. Perist. ext. t. narrowly lanceolate, slender, hyaline, with a sinuous dorsal line, int. processes from a broad carinato-plicate basal membrane.

Tribe xxix. LESKEACEÆ.

Pl. low, creeping, rarely erect, irregularly or pinnately branched. L. patent or secund, papillose, nerved, opaque, upper cells minute, hexagonal or dot-like, lower generally lax and hyaline; paraphyllia numerous. Calyptra cucullate, naked. Caps. erect, regular, or cernuous, incurved. Perist. double.

Fam. 1. Leskeæ.

St. prostrate, creeping. Br. more or less erect. Caps. erect and subcernuous. Int. perist. more or less perfect, cilia absent, or present.

104. **Myurella,** *Schp.* Pl. small, erect. Br. julaceous with the imbricate leaves. L. ovate, very concave, obsoletely two-nerved, obtuse or apiculate, finely serrulate, cells rhomboid, quadrate and hexagono-rectangular at base. Caps. on a smooth seta, oval-oblong, lid large, convexo-conical, apiculate. Calyptra small, fugaceous. Perist. large for size of caps., ext. t. perfect, int. basal membrane broad, processes entire, shorter than the binate cilia. Dioicous.

105. **Leskea,** *Hedw.* St. creeping. Br. erect or spreading. St. and br. l. similar, ovate-lanceolate, generally patent, rarely secund, papillose, nerved. Caps. oblong and elongate-cylindric, upright or sub-arcuate. Monoicous or dioicous.

106. **Anomodon,** *Hook. and Tayl.* Primary st. creeping, with minute leaves. Br. ascending or erect, irregular or fasciculate, stoloniferous. L. of secondary st. and br. patent or secund, cells minute, densely papillose. Caps. oblong, elongate, or cylindric, regular, coriaceous, annulus narrow or absent. Perist. t. pale,

D 2

linear-lanceolate, int. basal membrane narrow, processes short, more or less irregular. Dioicous.

Fam. 2.   Pseudo-leskeæ.

Resembling *Leskeæ.*   Caps. cernuous, horizontal, short, thick walled.   Perist. t. solid, coloured, int. membrane broader, with cilia generally between the perfect processes.

107. **Pseudo-leskea**, *B. and S.*   Habit, leaves, and cell formation as in *Leskea;* paraphyllia more or less numerous.   Caps. short, turgid.

Fam. 3.   Thuidiæ.

Pl. creeping and branched or erect and simple.   Br. pinnate, bi- tripinnate.   St. l. larger than br. l., obcordate-triangular, nerved, sulcate, cells upper small, rotund-hexagonal, lower elongate, in br. l. uniform, papillose on both sides; paraphyllia numerous.   Caps. cernuous, incurved.   Perist. t. long, densely articulate, confluent at base, int. membrane broad, processes long, with equally long cilia.

108. **Heterocladium**, *B. and S.*   St. creeping, rather rigid.   Br. subpinnate.   St. l. obcordate-lanceolate, shortly and obsoletely two-nerved; br. l. much smaller, ovate-acuminate, all serrate; per. l. squarrose.   Caps. oval or oblong, incurved, lid obtuse or rostrate, annulate.

109. **Thuidium**, *Schp.*   Primary st. prostrate, densely radiculose, or ascending and erect, with few radicles. Br. simple or bi- tripinnate.   St. l. largest, decurrent, obcordate-triangular, long or shortly acuminate, nerve solid, underneath or on both sides strongly papillose; paraphyllia various; br. l. smaller, ovate-lanceolate,

concave, imbricate; cells minutely rotund-hexagonal, at middle base generally linear-oblong, quadrate at sides, rarely all linear-hexagonal; per. l. numerous, elongate. Caps. on a long seta, oval, oblong, cylindric, subarcuate, cernuous and horizontal, lid with a long or short beak, annulus broad, rarely absent. Perist. ext. t. long, from a solid basal lamina densely articulate, reddish-brown, int. membrane broad, processes long, with 3 or 4 filiform cilia. Monoicous or dioicous.

Tribe xxx. HYPNACEÆ.

Very variable in size and habit. L. patent, squarrose, complanate, secund or falcate-secund, soft or scariose, shining or opaque, smooth, rarely papillose, nerved or nerveless, cells mostly narrowly prosenchymatose, often linear-serpentine, rarely parenchymatose. Calyptra cucullate. Caps. erect regular, or cernuous incurved, lid conical or rostrate. Perist. double.

Fam. 1. Pterygynandriæ.

Primary st. creeping, secondary arcuate, fertile. Br. fasciculate. L. apical cells rhomboid, middle narrowly rectangular or linear, basal angles and margin minutely quadrate; paraphyllia numerous, minute. Caps. erect, oblong-cylindric, lid conicorostrate. Perist. small, ext. t. pale, int. a narrow basal membrane with short processes and no cilia.

110. **Pterygynandrium,** *Hedw.* Filiform. L. patent or subsecund, nerved halfway, with long papillæ on back. Perist. short, processes imperfect.

Fam. 2. Orthotheciæ.

St. and br. prostrate, or dendroid from subterranean

stolons. L. patent or more or less complanate, ovate
or elongate-lanceolate, shining, nerved or nerveless,
cells densely linear, minutely quadrate at basal angles.

111. **Lescuræa,** *Schp.* Primary st. creeping, fertile
br. ascending, fasciculate. L. patent-erect, nerved,
smooth, sulcate. Perist. t. confluent at base, narrowly
lanceolate, int. processes scarcely as long as teeth,
irregular, appendiculate.

112. **Cylindrothecium,** *Schp.* Broadly cæspitose, low
or erect. Br. more or less distinctly pinnate. L.
complanate or patent. Calyptra as long as caps.,
which is elongate-cylindrical, lid conical. Perist. t.
linear-lanceolate, processes carinate, free almost or
quite to base.

113. **Climacium,** *W. and M.* Dendroid from a subterranean rhizome. St. l. lower squamiform, colourless, upper erecto-patent, decurrent, sulcate, thinly
nerved, oblong-lanceolate or ovate. Calyptra dimidiate. Caps. erect, oblong-cylindric, lid beaked.
Annulus absent. Perist. large, ext. t. linear-lanceolate,
with a serpentine dorsal line, processes from a narrow
basal membrane, perforate.

114. **Isothecium,** *Brid.* Dendroid from a creeping
primary st. L. patent, broadly ovate or ovate-oblong,
more or less suddenly acuminate, concave, nerved
about halfway, cells linear, at base quadrate. Caps.
erect, oblong-cylindric, lid conical, with a thick beak.
Perist. int. processes entire, cilia imperfect. Dioicous.

115. **Orthothecium,** *Schp.* Slender, prostrate, and
irregularly branched, or erect and br. fastigiate. L.
subsecund or erecto-patent, elongate-lanceolate, acuminate, entire, nerveless, sulcate, basal cells scarcely
quadrate. Calyptra small, fugaceous. Caps. erect,

oval or oblong, straight or curved, lid convex, with a short beak. Annulus broad. Perist. t. lanceolate-subulate, thin, hyaline, processes from a basal membrane as long as or longer than teeth, with or without cilia.

116. **Pylaisia**, *Schp.* St. creeping or erect, pinnate. Br. short. L. patent or secund, silky. Calyptra long-beaked, half-length of caps., which is oval, oblong, and subcylindric, often slightly curved. Perist. t. densely articulate, solid, processes from a narrow basilar membrane, as long as teeth, linear-subulate, carinate or split.

117. **Homalothecium**, *Schp.* Robust, partly creeping, partly erect, irregularly branched. L. silky, sulcate, thinly nerved, cells narrowly linear. Calyptra large, sometimes slightly hairy. Caps. erect, on a smooth or papillose seta, oblong-elongate, straight or slightly curved, lid with a short beak. Perist. int. membrane narrow, processes slender, shorter than teeth, with or without cilia.

Fam. 3. **Camptotheciæ.**

L. patent, scariose, shining, nerved, sulcate, cells linear or vermicular, quadrate at angles. Caps. oblong, cernuous, when dry arcuate, horizontal. Int. perist. processes and cilia perfect.

118. **Camptothecium**, *Schp.* Habit and leaves of 117. Caps. incurved, cernuous, on a rough or smooth seta, lid shortly rostrate or conical. Int. perist. a broad basal membrane, processes with long cilia.

Fam. 4. **Brachytheciæ.**

Generally prostrate, irregularly branched. L. generally patent, rarely secund, subcordate-ovate,

more or less acuminate, cells narrowly hexagono-
rhomboid, minutely quadrate at angles. Caps. on a
smooth or rough seta, cernuous, short, thick, turgid,
lid tumid, conical, with a long or short beak.

119. **Ptychodium,** *Schp.* Tufts entangled. Br. subpinnate. L. patent or secund, with several deep
plicæ, slenderly nerved to apex, ovate-lanceolate, cells
linear-rhomboid. Perist. ext. t. long, int. membrane
with processes as long as teeth, with rudimentary
cilia.

120. **Brachythecium,** *Schp.* Pl. irregularly branched.
L. patent, rarely subsecund, silky, more or less decurrent at angles, thinly nerved, irregularly sulcate,
cells narrowly hexagono-rhomboid, generally quadrate
at base. Caps. cernuous, on a rough or smooth seta,
turgidly ovate or subglobose, rarely oblong, lid conical.

121. **Myurium,** *Schp.* L. densely imbricate, julaceous, very concave, cochleariform, suddenly passing
into a long filiform apiculus, shining, nerve absent or
slender, cells vermicular. Fruct. unknown.

122. **Scleropodium,** *Schp.* General habit of *Brachythecium*, with l. cells narrow, vermicular, at decurrent
angles excavate, hyaline. Caps. suberect or cernuous,
oblong or ovate, more or less incurved, seta rough.
Perist. inner processes with appendiculate ciliolæ.
Dioicous.

123. **Hyocomium,** *Schp.* Br. pinnate, densely leafy.
St. l. broadly triangular or cordate, decurrent, with a
long apiculus, serrulate, nerve short, bipartite, upper
cells flexuose-linear, middle linear-rectangular, at angles
hexagono-rectangular. Caps. on a papillose seta,
turgid, oval, lid convex, apiculate.

124. **Eurhynchium,** *Schp.* Pl. small or taller. Br.

more or less distinctly pinnate. L. patent, rarely subsecund, from a narrow decurrent base, obcordate-or broadly ovate-lanceolate, serrate, nerved, smooth, cells narrowly rhomboid-hexagonal and subvermicular, dilated at basal angles. Seta rough or smooth. Caps. cernuous or horizontal, turgidly oval or oblong-incurved, lid more or less long-beaked. Dioicous or rarely monoicous.

125. **Rhynchostegium,** *Schp.* Minute, sometimes larger, creeping, irregularly branched. L. patent or complanate, those of st. and br. similar, lanceolate, ovate or oblong acuminate, nerveless or single nerved, shining, cells narrowly rhomboid. Monoicous or dioicous. Caps. cernuous or horizontal, ovate, lid subuli-rostrate, seta smooth, rarely rough.

Fam. 5. **Hypnæ.**

Pl. creeping, or erect. L. smooth, either patent, erect, or imbricate, secund or falcato-secund, cells generally narrow, often vermicular, rarely lax and prosenchymatose or partly parenchymatose. Caps. from a short neck, erect, cernuous or horizontal. Perist. t. linear-lanceolate, more or less subulate, densely articulate, int. membrane with long processes, usually with cilia.

126. **Plagiothecium,** *Schp.* Pl. prostrate or partly ascending, irregularly branched, stoloniferous, rooting. L. complanate, rarely turned to one side, tender, shining, nerve absent or obsoletely double, cells long, hexagono-rhomboid. Fr. springing from base of branches. Caps. suberect or cernuous, more or less curved, thin-walled, lid convexo-conical or rostrate. Perist. cilia with or without appendages.

127. **Amblystegium,** *Schp.* Pl. small and ascending, irregularly branched. L. soft, singly nerved, rarely nerveless, cells partly prosenchymatose, partly parenchymatose, lax. Caps. on a smooth seta, suberect or cernuous, oval or arcuate-cylindrical, lid large, tumid, conical, obtuse. Perist. int. processes usually entire, ciliola more or less perfect, rarely absent.

128. **Hypnum,** *Dill.* Pl. of diverse habit, low or erect, with or without radicles. Br. irregular or pinnate. L. patent or squarrose, often falcato-secund, nerveless, or singly or doubly nerved, shining or subscariose, cells narrowly linear or vermicular, at angles generally dilated, excavate. Caps. on a smooth seta, curved, cernuous, lid convexo-conical, mamillate, rarely rostrate. Perist. perfect.

129. **Hylocomium,** *Schp.* St. woody, pinnate or bipinnate, or sparingly branched. L. scariose, shining, sulcate, thinly two-nerved, without radicles, cells narrow, linear, broader at base. Caps. ovate or ovate-globose, coriaceous, lid mammillate. Perist. large, perfect.

Division I. SACCOMITRIA (Musci spurii).

Order I. *HOLOCARPÆ.*

Tribe i. ARCHIDIACÆ.

1. ARCHIDIUM, *Bridel.*

1. A. alternifolium, *Dicks,* Braith. Br. M. Fl., p. 92, t. 14a (*A. phascoides,* Wils. Bry. Brit., Syn. i. 26). St. ¼ inch, second year branched, sometimes 1 inch. Fertile branches short, barren ones longer, slender, and with more distant leaves. L. lanceolate pointed, upper ones longest, entire, nerved nearly to or beyond apex; per. l. ovate-lanceolate, toothed near the apex, nerve excurrent.

Moist clayey or chalky banks, &c. March, April.

Order II. *SCHIZOCARPÆ.*

Tribe i. ANDREÆCEÆ.

2. ANDREÆA, *Ehr.*

a. L. nerveless.

2. A. petrophila, *Ehr.,* Braith. Br. M. Fl., p. 6, t. 1a, and Edition 1 (*A. rupestris,* Hedw., Bry. Brit.). St. ¼ in., loosely tufted, with fastigiate branches. L. erect, imbricate, with a sheathing base, generally secund, ovate or ovate-lanceolate, papillose entire, tapering above, and rather obtuse; per. l. larger, convolute; all reddish-brown.

Subalpine rocks, frequent. May, June.

Var. β. HOMOMALLA, *Thed.* St. short, laxly pulvinate, olivaceous green above, fuscous below. L. ovate or oblong lanceolate, obtuse, more or less falcato-secund. Braith.

Glen Callater, Braemar; Castel-y-Gwynt, 3000 feet, Carnarvon.

Var. γ. ACUMINATA, *Schp.* More robust, olive-green, or blackish. L. longer, more acuminate, papillæ longer. Br.

Glen Callater, Ben McDhui, Cader Idris, Strachan, Kincardineshire; Shetland; Ben Nevis, 4000 feet, Abergynalwyn, &c.

Var. δ. FLACCIDA, *Schp.* Soft black tufts, st. branched flexuose. L. squarroso-patent, lanceolate, pointed, rather obtuse. Br.

Glen Callater; Canlochan.

Var. ε. SYLVICOLA, *Schp.* Short, lax, slender. L. small, acute, longly lanceolate-acuminate. Br.

Ben McDhui, Glen Callater; Loch Kandor.

Var. ζ. GRACILIS, *Schp.* Rufous brown or rosy purple. L. suberect, broadly oblongo-lanceolate.

Stye-head Pass; Cader Idris; Loch-na-gar, Ben Nevis.

Var. η. ALPESTRIS, *Thed.* Black-brown, slender, much branched. L. small, obtuse, laxly areolate, less papillose. Br.

Glen Callater, Morone, Braemar; Ben Challum, Uam Mohr. (*A. alpestris*, ed. 1, p. 21.)

Var. θ. SPARSIFOLIA, *Lind.* Small, lax, rufescent, slender, flexuose. L. small, spreading, lanceolate, gradually acuminate. Br.

Summit of Ben More, Perthshire.

3. **A. alpina,** *Smith,* Braith. M. Fl., p. 10, t. 1b; ed. 1.

p. 21.   St. tall, 1—3 inches, densely tufted, with long fastigiate branches. L. imbricate, spreading, chocolate-brown, obovate, almost panduriform, apex acute, denticulate at base; per. l. larger, elliptic, sheathing, ovate-oblong acute, convolute.

Alpine rocks.   May, June.

Var. β. COMPACTA, *Hook.*   Tufts dense, blackish-purple. Br. equal. L. closely imbricate. Br.

Ben Nevis; Great Glydr, N. Wales.

Var. γ. FLAVICANS, *Hook.*   Elongate, filiform. L. yellowish, loosely imbricate. Br.

Summit of Ben Nevis.

4. **A. nivalis,** *Hook.*, Braith. M. Fl. p. 15, t. 2*b.*   St. long, tufted, slender. L. reddish-brown, falcato-secund, lanceolate-subulate, acute, gradually tapering and nerved to apex, densely papillose on both sides. Dioicous.

Alpine rocks.   July, August.

Ben Nevis, Ben Cruachan, and Ben McDhui.

Var. β. FUSCESCENS, *Hook.*   More flexuose. L. more strongly falcate.

Ben Nevis; Ben McDhui.

b. L. nerved.

5. **A. Rothii,** *W. and M.,* Braith. M. Fl., p. 10, t. 2*a* (*A. rupestris,* ed. 1, p. 22).   St. scarcely ½ inch, loosely tufted, almost black. L. imbricate, falcato-secund, longly subulate from an ovate base, nerve thin, flat, vanishing at apex; per. l. larger, inner convolute, nerveless.

Alpine and subalpine rocks.   June, July.

Devon and Cornwall; Yorkshire; Cumberland; N. Wales; Scotland; Cromagloun, Ireland.

Var. β. FRIGIDA, *Hueb.* (*A. Grimsulana,* ed. 1, p. 22). Larger, more robust, flexuose, prostrate. L. broader, more solid. Br.

Ben McDhui; Beamsley Fell, Yorkshire.

Var. γ. HAMATA, *Lindb.* L. lax, fuscous, green when young, strongly falcate, gradually tapering from base. Br.

Luggielow, co. Wicklow; Carfury, near Penzance.

Var. δ. FALCATA, *Schp.* (*A. falcata,* ed. 1, p. 22). Slender. L. falcato-secund, suddenly lanceolate-subulate from a dilated obovate base, nerve flattened, ending at or below apex, which is slightly erose at margin.

Alpine rocks, not uncommon.

Scotland; Cumberland; Devon; N. Wales, &c.

6. **A. crassinervis,** *Bruch.*, Braith. M. Fl., p. 11, t. 1c. Tufts depressed, deep black. St. prostrate, ascending, fragile. L. shining, falcato-secund, subulate from an oblong base, nerve thick, excurrent into the round papillose subula, margin entire, cells quadrate; per. l. erecto-patent, convolute, nerveless.

Alpine rocks.   July, August.

Hebden Bridge, 1864; Scotland; Snowdon, 1853; Cumberland; Cheshire; N. Wales; Upper Lough Bray, Ireland.

Tribe ii. SPHAGNACEÆ.
3. **SPHAGNUM,** *Dill.*

[N.B.—The arrangement of this genus is adopted from Dr. Braithwaite's splendid Monograph, and the diagnoses are, in the greater part, abbreviated from his, as indeed they could not be better described.]

A. Cymbifolia.

Plants robust, loosely tufted. Br. turgid, those of

the coma obtuse. Br. l. imbricated, very broad, ovate, rounded and cucullate at apex, boat-shaped, concave, scarcely margined. Dioicous.

a. Cells of bark of stem with spiral fibres and pores.

7. **S. cymbifolium,** *Ehr.* St. 3—12 inches, robust, tufted, solid, covered with a cortical web. Tufts dirty olive-green, sometimes purplish. Bark of stem of 3 or 4 strata of cells, both fibrose and porose. St. l. lingulate-spathulate, with a rounded, slightly frayed apex, usually without fibres or pores ; br. l. imbricate, broadly ovate, concave, cucullate and muriculate at apex. Chlorophyll cells entirely enclosed by the hyaline cells. Caps. large, globose, on a short seta. Dioicous.

Bogs, common. June, July.

Var. *β.* SQUARROSULUM. Slender, dark green. Br. l. more pointed, slightly squarrose.

Var. *γ.* COMPACTUM, *Schultz* (*congestum,* Schp.). Densely tufted. St. green or purplish. Br. l. suberect, obtuse ; st. l. upper cells fibrose.

b. Cells of bark of stem porose but not fibrose.

8. **S. papillosum,** *Lindb.* Tufts ochraceous, never tinged with purple. St. l. spathulate, fringed at apex, basal cells large, empty, upper with a few fibres and pores. Br. l. broadly ovate obtuse, cucullate, concave, papillose at apex; chl. cells enclosed in hyaline, ellip-tical, and the latter covered internally, where uniting with former, with fine papillæ; lower cells of pedun-cular bracts, in centre thick-walled, empty ; marginal and upper cells porose and fibrose.

Subalpine peat bogs, frequent.

Var. *β.* CONFERTUM. Tufts short, dense. Bark cells of st. usually non-fibrose. Br. l. very concave, obtuse.

Near Penzance.

Var. γ. STENOPHYLLUM. Br. l. ovate-oblong, less concave and almost entire above.

Penzance; Staveley.

c. Cells of bark of stem, outer without fibres, inner with fibres and large pores.

9. S. Austini, *Sull.* Ochraceous. St. dark brown, tufts very thick and spongy. St. l. lingulate, obtuse, fringed, cells as in No. 7. Br. l. closely imbricate, ovate-oblong, concave, less cucullate, strongly muriculate at back, cells with fibres and large pores; chl. cells triangular, projecting between the hyaline on concave surface of leaf, the latter at union densely papillose internally. Peduncular bracts with lower cells narrow, empty, upper fibrose and porose.

Grows in great elevated hassocks in swamps.

Isle of Lewis, 1868; Moss Raplock, Barend Moss, and Auchencairn Moss, Kirkcudbright (J. McAndrew, 1882). These are the only known British localities.

Var. β. FIMBRIATUM, *Braith.*, Sphag., p. 34 (*imbricatum*, Hornsch). Dense-cushioned, dark brown tufts. Br. short, densely crowded. L. closely imbricated.

Isle of Lewis (Moore), Witherslack Moss (Barnes), Moss Raplock (McAndrew, 1882).

B. Subsecunda.

Soft, loosely tufted. Br. l. usually subsecund, erecto-patent, broadly ovate, concave, broadly margined, margins involute above. Dioicous.

10. S. tenellum, *Ehr.* Pale greenish-yellow. St. short, straw-coloured, bark of two layers not porose. Br. pale red, loose leaved, retort-cells of bark with

apex recurved, projecting, perforated. St. l. large,
reflexed, ovate-oblong, margin incurved, cells densely
fibrose with a few pores above, near base empty. Br. l.
broadly ovate, elongato-lanceolate, margin incurved in
upper half, margined; chl. cells triangular on back of
leaf, interposed between the hyaline, which are strongly
fibrose. Male amentula orange-coloured.

Spongy heaths, &c., in hilly places.　　May, June.

Lancashire, Yorkshire, Sussex, Kent, Scotland,
Ireland.

11. **S. rubellum**, *Wils.* St. 2—5 inches, slender,
purple, with slender, deflexed, sometimes curved
branches, bark cells destitute of pores; st. l. large,
erect, ovate-oblong, concave, obtuse, subsecund, with
a minutely toothed apex and narrow margin, cells
without fibres or pores. Br. l. ovate or oblong-ovate,
gradually acuminate, apex with three stout teeth and
a few smaller ones below, cells with annular fibres and
a few pores, becoming narrower towards the edge, chl.
cells triangular, interposed on upper surface. Caps.
almost included.

Bogs among grass and carices, fr. rare. June, July.

12. **S. laricinum**, *Spruce* (*neglectum*, Angst.). St.
4—6 inches, pale yellowish-brown, bark pale of 2—3
layers of cells. Br. crowded 3—4, of which 1—2 are
divergent, the others pendent but not appressed. St.
l. small, ovate, cucullate, minutely fimbriate, basal cells
utricular, middle and lateral very narrow, apical,
rhomboid, with scarcely any fibres or pores. Br. l.
ovate, shortly and bluntly acuminate, point with 3—5
teeth, recurved, broadly bordered, cells with annular
fibres and many pores, chl. cells elliptic, central
between the hyaline. Ped. bracts oblong, scarcely

E

bordered.   Retort cells of branches without any projecting neck.

Deep bogs; no doubt often mistaken for *subsecundum.*

Near Holyhead; Vale Royal, Cheshire; Loch Kander,
Braemar.   First found by Dr. Spruce at Terrington
Carr, Yorkshire, in 1846.

Var. γ. PLATYPHYLLUM.   Br. short. St. l. lingulate,
distinctly auricled; br. l. very broad and concave.

Pass between Aber and Llanrwst, N. Wales.

Var. δ. CYCLOPHYLLUM.   Tufts generally short, turgid,
simple.   St. l. very large, orbicular, concave, cucullate.

Shore of Loch Katrine.

13. S. subsecundum, *Nees.*   Tall and slender.   St.
brown or blackish in the type (green in vars. β and γ),
with only one layer of cortical cells.   Br. 2—3,
spreading, 1—2 pendent but not appressed, retort cells
slightly recurved at apex and perforate.   St. l. small
(large in var. δ), more or less auricled, cucullate, very
minutely fringed, border narrow, upper cells fibrose
and porose, lower almost without fibres.   Br. l. more
or less subsecund, broadly acuminato-elliptic, very
concave, narrowly bordered, apex 3—5 toothed, hyaline
cells elongate, fibrose, numerously porose, chl. cells
central, much compressed.

Turf bogs, and about springs and streams in moorlands, &c.   Frequent.   June, July.

Var. β. CONTORTUM.   Dark olive-green.   St. green.
Br. crowded, usually more or less contorted and
twisted.   Br. l. larger, broader.

Deep bogs.   Frequent.

Var. γ. TURGIDUM, *Muell.*   St. green.   Br. swollen,
with large very broad leaves, truncate at apex,

5-toothed. St. l. large, fibrose in upper cells, or sometimes throughout.

Ditches and sides of pools. Not common.

Var. δ. AURICULATUM. St. generally greenish; st. l.
large, lingulato-acuminate, subhastate at base, with
very large auricles, composed of fibrose utricular cells,
apex truncate and erose.

Hayward's Heath, Sussex, &c.

C. Rigida.

Densely ramulose, forming compact, cushion-like
tufts. Br. l. erecto-patent, oblong, concave, very
narrowly bordered, apex obtuse, truncate, and toothed,
margin involute for the greater part of its extent.
Monoicous.

14. **S. rigidum**, *Schimp*. Tufts dense, rigid when
dry. St. 3—10 inches, dark brown, closely branched,
bark cells small, non-porose, of 2—3 strata. Br. 3—4,
short, 1—2 erecto-patent, obtuse, the rest slender,
deflexed, flagelliform. St. l. minute, erect, oblique,
broadly ovate, auricled, apex erose, border broad, cells
without fibres or pores, or a few fibres only near base.
Br. l. ovate-oblong, slightly cucullate, cells fibrose,
with unequal pores, chl. cells small, much compressed,
central. L. of pendent br. narrower, distant.

Marshy heaths and moorlands. July.

Var. β. COMPACTUM. St. ½—2 inches. Br. short,
thick, and compact, generally pale tinged with red;
br. l. rounded at apex.

Drier places.

Var. γ. SQUARROSULUM, *Russ*. Tufts looser. Br.
more distant. L. of divergent br. squarrose.

15. [**S. molle**, *Sull.*] Tufts soft, whitish-green above,

pale brown below. St. 2—6 inches, slender, pale green, cortical cells in 2—3 layers, non-porose. Br. crowded, 2—3 in fasc., erecto-patent. St. l. very large, minutely auricled, ovate-spathulate, cells almost non-fibrose *in the type,* apex with 3 teeth and a few minute ones at sides. Br. l. oblong-ovate, concave, very narrowly bordered, apex truncate, with 5 or 6 irregular teeth, hyaline cells fibrose and porose, chl. cells triangular, projecting on concave surface of leaf.

Type not found in Britain.

Var. β. MÜLLERI. Br. 3—4 in fasc., 1—2 patulous, the rest longer, slender, pendent. St. l. more elongate, cells *with fibres and pores.* Perichætial bracts lanceolate acuminate, upper cells with fibres and pores.

Moorland streams. August.

Darnholme, near Whitby; Brickhill Heath, Bucks; Ben Lawers; New Forest, 1880 (H. Boswell).

Var. γ. TENERUM, *Braith.*, Sphag., p. 55. Tufts dense, dirty white. Br. closely crowded with acuminate leaves.

Darnholme (Braithwaite), Dalfroo Bog, Strachan (Sim), New Galloway, 1883.

D. Cuspidata.

Rather rigid, loosely matted. Br. l. erecto-patent, narrowly lanceolate, acute or much acuminate, truncate, toothed and involute at apex, margin more or less bordered.

a. Leaves erecto-patent.

16. S. acutifolium, *Ehr.* Tufts soft, pale green, often tinged with purple. St. 3—6 inches, with slender attenuated branches, cortical cells generally without pores, in 3—4 strata, woody zone purple, except in

var. *θ patulum*. St. l. small, ovate-acute, erect, concave, minutely auricled, apex 5-toothed, lower cells
without fibres or pores, upper often slightly fibrillose,
border broad, narrowing towards apex. Br. l. ovate-
lanceolate, slightly præmorse, with a 3—4 toothed apex,
hyaline cells with a few large pores and annular fibres,
chl. cells triangular on concave surface of leaf, border
extremely narrow, of two rows of cells. Pedicel long.
Monoicous.

Bogs and marshes. July.

* Varieties more or less rosy or purple.

β. DEFLEXUM. Br. all decurved, flagelliform, pink
and green, white-tipped; br. l. longer and narrower.

γ. LILACINUM. Fine rosy red suffused with violet.
Br. erect and divergent; br. l. strongly involute. St.
l. rounded at apex.

Terrington Carr and Westmoreland.

δ. PURPUREUM. Usually entirely purple. Br. short,
curved. L. obtuse.

ε. TENELLUM. Slender, elongate, pale green and red.
Divergent br. arcuate. L. short, obtuse; st. l. broadly
bordered, fringed at apex.

** Varieties of a fuscous or pale colour.

ζ. FUSCUM. Ochraceous brown. Divergent br. short,
incurved. L. short, concave, apex rounded, toothed;
st. l. small, apex rounded, toothed.

η. LURIDUM. Dirty green, fuscous below. Br.
crowded, all erecto-patent. L. acuminate, strongly
involute at apex; st. l. large, linear, acutely pointed.

Ben Lawers.

θ. PATULUM. Pale green. St. l. elongate, acute,
upper cells fibrillose. Br. l. elongate, patent.

ι. ARCTUM, *Br.* Fragile, yellowish, dense. St. l. narrowly bordered, cells of upper two-thirds fibrillose. Br. dense, short, ascending. L. obtuse, laxly areolate.

Brandon Mount, Kerry.

17. **S. fimbriatum,** *Wils.* Tufts pale whitish-green. St. slender, 6—12 inches, loosely cæspitose, bark cells 2—3, layers porose. Br. 3—4 in fasc., very long, attenuated, 2 arcuate and decurved, rest pendulous, filiform. St. l. obovate, broad, very obtuse, and fringed at the summit, cells without fibres or pores, border very broad below, but quickly narrows and vanishes; br. l. lower ovate-lanceolate, upper elongate-lanceolate, acute, cells fibrillose and porose, chl. cells compressed, enclosed, but nearer the upper surface; per. l. very large, obtuse, cucullate. Caps. on a short pedicel, nearly enclosed in the per. leaves. Monoicous.

Bogs and marshes. Frequent. June, July.

18. **S. strictum,** *Lindb. MS.* Resembles *fimbriatum,* but is more robust, yellowish-green or pale brownish. St. straight, pale, 6—10 inches, cortical cells 3—4, layers porose. St. l. erect, appressed, ligulate-spathulate, truncate, and laciniate-fimbriate at apex, cells without fibres or pores, with a broad border of narrow cells extending to apex. Br. 3—4, of which 2—3 are spreading, flagelliform, the rest deflexed, appressed, filiform. Br. l. erecto-patent, ovate-lanceolate and lanceolate, cells fibrose with numerous large pores, chl. cells trigonous, compressed near upper surface. Dioicous.

Shallow bogs on subalpine heaths.

Ben Ledi, Killin, Ben Lawers, Glen Lyon, Banchory; Dent and Skegglesmere in Westmoreland.

Var. β. squarrosulum, *Russ.* Plants very small. Br. l. recurved at apex.

19. **S. squarrosum,** *Pers.* St. 6—15 inches, rigid, reddish-brown, often forked, cortical cells in 2 layers, non-porose. St. l. large, lingulate, not bordered, minutely auricled, apex rounded, slightly fimbriate, cells without fibres or pores. Br. 4—5 in fasc., 2—3 divergent, tumid, attenuated towards apex. L. on lower two-thirds squarrose and recurved from middle, rest elongate; leaves of pendulous branches appressed, slender, imbricate. Br. l. base broadly ovate, concave, suddenly lanceolate, apex minutely 3—4 toothed, margin of 2—3 rows of very narrow cells, chl. cells compressed, entirely enclosed by hyaline, which are fibrose and porose. Monoicous.

Boggy places near moorland streams, &c. Frequent. July.

Var. β. squarrosulum. Smaller. St. pale green. L. more distant, small.

More shady alpine places.

20. **S. teres,** *Angst.* Tufts small, soft, pale yellowish-green, often ferruginous. St. slender, 4—8 inches, pale red, cortical cells 3 layers, non-porose. St. l. as in *squarrosum.* Br. distant, 4—5 in fasc., 2—3 divergent, terete. L. imbricate throughout, their apices only slightly recurved, broadly ovate, pointed, 3-toothed. Dioicous.

Edges of subalpine bogs and springs. Not common.

Knutsford Moor, Wybunbury Bog, Newchurch Bog, all in Cheshire; Dorme, in Scotland; Staveley, Westmoreland.

21. **S. Lindbergii,** *Schp.* Tufts dense, 6—12 inches, yellowish-green tinged with red or purplish-brown.

St. solid, dark brown, cortical cells 3—4 layers, irregular, without pores.   St. l. crowded, reflexed, broadly lingulate, auricled, apex broad, truncate, fringed, fibres and pores occur sparingly in the large auricles, other cells generally empty.   Fasc. of 3—4 branches, 2—3 arcuate and divergent, the rest pendent, elongate and closely appressed.   Br. l. in 5 rows, not undulate, firm, glossy, ovate at base, lanceolate above, toothed and involute at apex, cells fibrose and porose, chl. cells narrow, elliptic, quite enclosed but nearer back of leaf, border widest at base of 3—4 rows of very narrow cells.   Perichætium large, inflated.   Monoicous.

Deep bogs.   July.

Ben Wyvis, Ross-shire (McKinlay, 1867).

22. S. intermedium, *Hoffm.*   Pl. robust, straight, yellow or pale green, pale brown or whitish below. St. 6—12 inches, greenish-white, cortical cells 2—3 layers, small, non-porose.   St. l. small, reflexed, ovato-triangular, minutely auricled without fibres or pores, broadly bordered, apex obtuse, 3—5 toothed.   Br. 4—5 in fasc., 2 arched downwards, the rest pendent, closely appressed and concealing the stem.   Br. l. erecto-patent, densely imbricate, involute above, margins involute and points recurved when dry, apex minutely toothed, border narrow of 2—3 rows of very narrow cells.   Hyaline cells densely fibrose, with a few small pores, chl. cells trigonous, free on back of leaf. Dioicous.

Moorlands, wet heaths, and mountain bogs.   Frequent.   July.

Var. β. RIPARIUM.   Taller.   St. l. triangular, erose, fringed.   Br. l. ovate-lanceolate, non-fibrose at apex.

Var. γ. SPECIOSUM.   10—18 inches, deep green.   St.

l. large, longer, deeply fringed. Br. l. long, lanceolate, with a subulate point, recurved when dry, no fibres in upper cells.

Var. δ. PULCHRUM, *Braith.*, Sphag., p. 81. Robust, densely tufted, yellow-green, with a golden fulvous tinge above. St. l. acute, contracted at point into a minute, recurved apiculus, cells in upper part fibrillose.

Fowlshaw Moss and Broadgate Bog (Stabler), Wheeldale Moor (Anderson), Carrington Moss (Hunt), Whitchurch, Salop (Boswell, 1882).

23. **S. cuspidatum,** *Ehr.* Tufts soft, submerged or floating, generally greenish. St. 6—18 inches, slender, pale green, cortical cells 2—3 layers, not porose. St. l. ovate-oblong, with a broad border of very narrow cells, upper half fibrose. Br. 3—5 in fasc., all divergent or 1—2 pendent but not concealing the stem. Br. l. narrowly lanceolate, imbricate, flexuose when dry, often slightly falcato-secund, with a broader border. Chl. cells trigono-elliptic, free on posterior surface. Dioicous.

Moorlands in pools. Frequent. July.

Var. δ. TORREYANUM. Very large and robust, colour dirty brown. St. stout. St. l. large, without fibres; br. l. very large, elongate, lanceolate-acuminate, broadly margined, erose-denticulate, and semitubular at apex.

Wet bogs; near Whitchurch, Salop. H. Boswell, October, 1882.

DIVISION II.   STEGOMITRIA (Musci genuini).

Section I.   *ACROCARPI.*

Tribe iv.   WEISSIACEÆ.

Fam. **Weissiæ.**

4. SYSTEGIUM, *Schp.*

24. **S. crispum,** *Hedw.* St. ¼ inch, cæspitose, with fastigiate branches. L. lanceolate-subulate, grooved, erect or spreading, margins involute, nerve excurrent; per. l very long, concave at base, sometimes almost secund. Caps. roundish, immersed, pale brown, beak oblique.

Banks and fields, chiefly limestone.   Spring.

25. **S. Mittenii,** *Schp.* (ed. 1, p. 31).   St. fragile, flexuose, erect. St. l. squarrose and recurved, broadly lanceolate, nerve thick, terete, vanishing at apex. Caps. on a longer pale seta, ovate, lid small, beak rostrate.

Damp clay soils, roadsides, &c.   Spring.
Hurstpierpoint, Sussex. Mr. Mitten.

26. **S. multicapsulare,** *Smith.* St. ½ in., loosely tufted. L. distant, alternate, spreading, lanceolate, recurved above the middle, somewhat obtuse, with an excurrent nerve and plane margin; per. l. longer and broader, erect, incurved. Caps. ovate, elliptic, tapering to an oblique short beak, on a longish pedicel, sometimes two together.

Clay fields, &c. Rare.   March.
Sussex, Cheshire, Yorkshire, &c.

5. HYMENOSTOMUM, *R. Br.*

27. **H. rostellatum,** *Brid.* St. $\frac{1}{8}$—$\frac{1}{4}$ inch, tufted. L. lower linear-lanceolate, upper longer, linear-lanceolate, erecto-patent, papillose, nerve excurrent, margin plane. Caps. olive-brown, ovoid, elliptical, with a straight beak, scarcely exserted, pedicel equalling caps. in length.

Dried beds of pools, &c. Autumn, spring.

28. **H. microstomum,** *Hedw.* St. $\frac{1}{8}$—$\frac{1}{4}$ inch, densely tufted. L. linear-lanceolate, acute, patent from an erect base, upper ones longest, nerve excurrent, mucronate. Caps. elliptical, exserted, sometimes oblique and gibbous, olive-brown, much contracted, lid with a longish curved beak.

Fields, &c., clay, sand. Spring.

Var. *β.* OBLIQUUM. Caps. oblong, oblique, lid subrostrate.

Var. *γ.* BREVIROSTRE. Caps. oblong, symmetrical, lid short, conical.

Var. *δ.* BRACHYCARPUM. Caps. roundish, gibbous.

Var. *ε.* ELATUM. Innovations overtopping fruit. Caps. roundish, small, lid subrostrate.

29. **H. commutatum,** *Mitt.* "L. from a suboblong base, lanceolate, narrowed, keeled with the nerve, which vanishes below apex, cells nearly all elongated and pellucid; per. l. similar." Caps. turbinate, lid with a very oblique longish beak.

Alpine rocks.

Nant-y-Fydd, Wrexham. Mr. Bowman.

30. **H. squarrosum,** *Nees and H.* St. $\frac{1}{4}$ inch, loosely tufted. L. linear-lanceolate, squarrose, distant, blunt, nerve running out into a mucro. Caps. elliptical, symmetrical, lid conical with a blunt beak.

Clayfields and banks.   Autumn, spring.

31. **H. tortile,** *Schw.*   St. ⅛—¼ inch, densely tufted
with fastigiate branches. L. oblong-lanceolate, spreading or suberect, curved, concave, margin incurved,
obtuse, pointed with the excurrent nerve. Caps. ovate-
oblong, with a purple mouth and an inclined beaked lid.

Limestone rocks.   May, June.

6. GYMNOSTOMUM, *Bry. Eur.*

32. **G. calcareum,** *Müll.*   St. short, simple, densely
tufted, radiculose at base. L. lower small, ferruginous,
erecto-patent, narrowly lanceolate, upper larger, deep
green, linear-lanceolate, rather obtuse, concave, stoutly
nerved nearly to apex, margin minutely crenulate;
per. l. lanceolate, concave, acute. Caps. oblong, subcylindric, short-necked, erect, on a pale yellow seta,
lid conical, subulate.

Cheedale, Derbyshire. W. West, 1880.

Var. δ. BREVIFOLIUM, *Schpr.* Slender branched. L.
lower very minute, distant, upper crowded, ovate-
lanceolate, recurved above. Caps. oval.

Damp rocks and walls.

Blackhall, near Banchory, Deeside. Mr. Sim.
Barren.

33. **G. rupestre,** *Schw.*   St. 1—2 inches, densely
tufted, slender, dichotomous. L. linear-lanceolate,
more or less papillose, spreading, obtuse, keeled, thickly
nerved nearly to apex. Caps. erect, oval, lid flattish,
suddenly rising to a longish scarcely bent beak.

Wet alpine rocks.   Autumn.

34. **G. curvirostrum,** *Hedw.*   St. ½—1 inch, cæspitose
branches fastigiate. L. erecto-patent, linear-lanceolate,
spreading keeled, margins recurved, smooth, or mi-

nutely papillose, nerved nearly to apex. Caps. broadly ovoid, lid adhering to columella, conical flattened, with a long suddenly bent beak.

Moist subalpine calcareous rocks.   Autumn.

7.   ANŒCTANGIUM, *B. and S.*

35. **A. compactum,** *Schleich.* St. 1—4 inches, densely tufted, slender. L. short, lanceolate, spreading from an erect base, acuminate, densely papillose, slightly serrulate, nerved to or beyond apex. Caps. oval-oblong, erect, mouth red, shining, lid long, convex, with a slender oblique beak.

Crevices of moist alpine schistose rocks.   Autumn.

8.   EUCLADIUM, *Br. and Sch.*

36. **E. verticillatum,** *L.* (?)   St. ¼—¾ inch.   Br. fastigiate. L. linear-lanceolate, rigid, denticulate at base, above remotely crenulate, suberect, margin plane, with a strong slightly excurrent nerve. Caps. erect, reddish, lid subulate.

Dripping limestone rocks.   June, July.

9.   GYROWEISSIA, *Schp.*

37. **G. tenuis,** *Schrad.* St. tufted. L. lingulate, suberect, upper ones longest, entire, obtuse, concave, nerved nearly to apex. Caps. oblong elliptic, pale brown, lid obtuse, shortly rostrate.

Sandstone rocks and walls.   July, August.

10.   WEISSIA, *Hedw.*

38. **W. viridula** (ed. 1, *W. controversa*).   St. ⅛—½ inch, branched. L. lower lanceolate, upper linear-

lanceolate, margin incurved, with a slightly excurrent nerve. Caps. oval, erect, lid conical, beak half-length of capsule, seta twisted, reddish. Barren fl. gemmiform.

Frequent. Spring.

39. **W. mucronata,** *B. and S.* Smaller than last. L. linear-lanceolate, with plane margins, concave above, the nerve more excurrent and forming a mucro. Caps. oblong, elliptic, teeth of per. short, truncate, perforated, lid with a longish beak. Barren fl. gemmiform.

Fallow (clay) ground. March, April.

11. DICRANOWEISSIA, *Lindb.*

40. **D. crispula,** *Hedw.* St. shorter than last, branched. L. spreading, frequently falcato-secund, lanceolate-subulate, base wide, concave, margins plane, not nerved to apex. Caps. oval or oblong without annulus, lid beaked. Barren fl. gemmiform.

Mountainous rocks. June, July.

41. **D. cirrhata,** *Hedw.* St. ½—1 inch, loosely tufted. L. linear-lanceolate, spreading, entire, concave, keeled, margin reflexed, not nerved to apex; per. l. slightly sheathing, shorter. Caps. oval-oblong, lid with a long beak.

Posts and rocks in mountainous districts. Spring.

42. **D. Bruntoni,** *B. and S.* (ed. 1, *Cynodontium*). St. ½—1 inch, tufted. Br. fastigiate. L. linear-lanceolate or lanc-subulate, flexuose, concave, minutely papillose, keeled, sometimes minutely denticulate at apex, margin reflexed, twisted when dry, nerved almost ox quite to apex; per. l. sheathing. Caps. erect, oval-oblong or elliptical, smooth, with a long oblique beak.

Subalpine siliceous rocks. May, June.

12. **RHABDOWEISSIA,** *Bruch. and S.*

43. **R. fugax,** *Hedw.* (*Oncophorus striatus,* Br. M. Fl., 172). St. ¼—½ inch, tufted. L. linear-lanceolate, acute, toothed near apex, margins plane. Caps. ovate, somewhat striated; teeth of per. subulate, fugacious, lid with an oblique beak longer than capsule.

Subalpine rocks, in crevices.   June, July.

44. **R. denticulata,** *Brid.* (*O. crispatus,* Br. M. Fl., p. 171). St. longer than last, loosely tufted. L. lingulate or linear-lanceolate, strongly toothed halfway from apex. Caps. more distinctly striated when dry, teeth of per. lanceolate, persistent.

Alpine and subalpine rocks.

Tribe v. DICRANACEÆ.

Fam. 1. **Pseudo-Dicranæ.**

13. **CYNODONTIUM,** *B. and S.*

45. **C. gracilescens,** *W. and M.* "L. patent, tortuose, lanceolate, scarcely acuminate, rather obtuse, densely papillose on both sides, nerved nearly to apex. Caps. on a flexuose seta, erect, oblong, not strumose" (*R. Braithwaite*), ribbed when dry, with a long-beaked lid.

Glen Phee, Clova. Very rare.   August,

46. **C. polycarpum,** *Ehr.* (ed. 1, *Dicranum; Oncophorus,* Br. M. Fl., 169). L. bent, flexuose, lanceolate-subulate or linear-lanceolate, keeled, margin recurved, somewhat papillose, denticulate at apex, nerve excurrent. Caps. erect, symmetrical, striated, neck tapering.

Alpine rocks.   July, August.

47. **C. strumiferum,** *Ehr.* Slender, branched. L. lanceolate, subulate, generally slightly denticulate at apex, where nerve vanishes, papillose. Caps. sub-

cernuous or cernuous, gibbose-ovate, distinctly strumose.

Glen Phee, Braemar, Teesdale, &c. Rare. August.

48. **C. virens,** *Hedw.* (ed. 1, *Dicranum; Oncophorus,*
Br. M. Fl., 165). St. 1—3 inches, branched, radiculose.
L. erect, ovate-lanceolate at base, sheathing, running
to a long subdenticulate, almost setaceous prolongation,
margins recurved, nerve thick, subexcurrent. Caps.
cernuous, strumose, smooth, oblong, and curved, lid
beaked.

Moist alpine rocks; Ben Lawers, &c.   June, July.

Var. β. SERRATUS. L. patent, spreading, coarsely
serrate.

" N. of England, Prof. Barker, 1876."—*Braithwaite.*

49. **C. Wahlenbergii,** *Brid.* (*Oncophorus,* Br. M. Fl.).
St. short, scarcely branched. L. distant, narrowed
suddenly from a broad sheathing base into a long
linear acute subula, entire, or sometimes faintly toothed
near apex. Caps. obovate, short, much incurved, only
slightly strumose.

Glen Callater. Rare.   August.

Var. β. COMPACTUS. Very compact. L. denser,
strongly curled, entire.

Braemar, Little Craigandal.

14. DICHODONTIUM, *Schp.*

50. **D. pellucidum,** *Hedw.* St. 1—2 inches, loosely
tufted. L. distant, lanceolate, squarrose, or patent,
spreading, margins undulate, denticulate at apex,
papillose, obtuse. Caps. shortly ovate, cernuous,
gibbose, lid conical, rostrate. Dioicous.

Wet stones in streams.   October, November.

Var. β. FAGIMONTANUM. St. short. Br. slender. L. shorter.

Var. γ. SERRATUM. L. crenato-serrate, with a more acute point. Caps. elliptic or oblong, erect or nearly so, lid with a slender beak.

Fam. 2. Dicranæ-veræ.

15. TREMATODON, *Rich.*

51. T. ambiguus, *Hedw.* St. branched, flexuose. L. erecto-patent, spreading, curved, from an ovate-oblong, concave base, lanceolate-subulate, channelled, nerve strong, excurrent into an apiculus. Caps. oblong, with a long tapering neck, subarcuate, on a flexuose yellow seta. Annulus broad, lid subulirostrate. Per. t. confluent at base into a membrane, unequally cleft, incurved when dry (Sch. Syn., p. 69).

Bare damp places, bogs, &c.   Summer.

Tummel Bridge, Perthshire. R. Braithwaite, M.D., August, 1883. In small quantity.

16. DICRANELLA, *Schp.*

Section 1. *L. squarrose or patent.*

a. Monoicous.

52. D. crispa, *Hedw.* St. ¼ inch, gregarious. L. subulate, from a broadish sheathing base, setaceous above, long, spreading, flexuose, minutely dentate, nerved to apex. Caps. almost erect, oval or obovate, striate, seta red, lid with a long, oblique, subulate beak. Monoicous.

Moist sandy banks. Not common.

August, September.

F

b. Dioicous.

53. D. Grevilliana, *B. and S.* (*Anisothecium Grevillei,*
Br. M. Fl.).   L. with a broad sheathing base, suddenly
lanceolate-subulate, prolonged, wide-spreading and
wavy, entire, nerve broad.   Caps. ovate, substriate,
strumose, lid with a beak longer than capsule.
Monoicous.

Damp clay, subalpine.   Rare.   September.
Glen Tilt, Blair Athole, Glen Shee.

54. D. Schreberi, *Hedw.* (*Anisothecium crispum,* Br.
M. Fl.).   St. ½ inch, subcæspitose, branched sparingly.
L. base broad, suddenly lanceolate-subulate, spreading,
flexuose, keeled, denticulate at apex.   Caps. ovate-
oblong, scarcely strumose, cernuous, not striate, lid
conical, shortly rostrate.   Dioicous.

Clayey or sandy soil near streams.   Rare.
                              October, November.
Lancashire, Cheshire, Yorkshire, Scotland, Ireland.

55. D. squarrosa, *Schrad.* (*Anisothecium,* Br. M. Fl.).
St. 1—3 inches, dichotomous.   L. lanceolate, from a
broad sheathing base, obtuse, undulate, entire, concave,
recurved, nerve narrow, reaching nearly to apex.   Caps.
ovate-oblong, cernuous, lid long, conical, with a short
beak.

Wet mountainous places.   August, September.

56. D. cerviculata, *Hed.*   St. ¼ inch, sparingly
branched.   L. spreading, flexuose, almost setaceous
from a broadish amplexicaul base, entire, nerved into the
subula.   Caps. roundish-ovate, gibbous, strumose, seta
yellow, lid with a long oblique or curved subulate beak.

Wet sandy banks or on turf.   Frequent.   June, July.

Var. β. PUSILLA.   St. shorter, simple.   L. smaller,
suberect.   Caps. smaller and less gibbous.

Section 2. *Secund or subsecund.*
a. Dioicous.

57. **D. varia,** *Hedw.* (*Anisothecium rubrum,* Br. M.
Fl.). St. ¼ inch, cæspitose. L. erecto-patent, lanceolate, entire, keeled, subdenticulate at apex, margin
reflexed, nerve scarcely excurrent. Caps.' inclined,
curved, ovate or oblong, slightly tumid, lid shortly
beaked, seta twisted to the right.

Moist banks. November, December.

Var. β. TENUIFOLIUM (*D. fallax,* ed. 1, p. 43). L.
narrow, obscurely nerved. Caps. nearly symmetric.

Var. γ. TENELLUM. St. slender, scarcely branched.
L. falcato-secund, distantly denticulate.

Var. δ. CALLISTOMUM. L. scarcely secund. Caps.
erect, obovate, truncated, lid almost as long as caps.

58. **D. rufescens,** *Turn.* (*Anisothecium,* Br. M. Fl.).
St. short, bright red, scarcely branched. L. erecto-
patent, linear-lanceolate, obscurely toothed, reddish,
margins plane, secund, pellucid, nerve strong, vanishing
at apex. Caps. erect, ovate or obovate, slightly tumid,
with a conical beaked lid, seta red, twisted to the left.

Moist clay banks. October, November.

59. **D. subulata,** *Hedw.* (*D. secunda,* Br. M. Fl.). St.
½—1 inch. L. falcato-secund, setaceous from an oblong,
lanceolate base, entire, nerve excurrent. Caps. ovate,
gibbous, cernuous, striate when dry, seta red; per. l.
sheathing.

Moist shady stony banks. September, October.

60. **D. curvata,** *Hedw.* Cæspitose. St. bi-tripartite.
L. setaceous from a shortly ovate semi-sheathing base,
channelled, apex denticulate, falcato-secund. Caps.
erect or suberect, ovate-oblong, slightly gibbous,
distinctly striate, seta pale red; per. l. not sheathing.

F 2

Wet sandstone rocks. Rare. October—February.
Llanberis, N. Wales (W. Wilson), Dolgelly, Dunoon.

61. **D. heteromalla,** *Hedw.* St. ½—1 inch, simple or
branched, in silky tufts. L. lanceolate, suddenly setaceous, slightly dentate at apex. Caps. obovate, suberect, tapering, slightly striate when dry, lid with a
long beak, seta pale yellowish.
Moist banks and walls. Very common.

November—March.

Var. β. STRICTA. L. erecto-patent, straight, not
secund, seta longer, flexuose.

Var. γ. INTERRUPTA. Larger. St. interrupted, leafy.
L. spreading or falcato-secund.

Var. δ. SERICEA, *Schp.* Plants taller. L. longer and
narrower, diverging almost on all sides, pale green or
yellowish, often strongly and remotely toothed.
Soccoth Hill, Arrochar (McKinlay).

<h3 align="center">17. DICRANUM, Hedw.</h3>

[I follow Dr. Braithwaite's arrangement entirely in
this genus.]

<h4 align="center">Section 1. ARCTOA.</h4>

Plants radiculose at base only. L. lanceolate, subulate, entire. Caps. small, with a tapering or strumose
neck.

62. **D. fulvellum,** *Sm.* St. ½—2 inches, densely
tufted. L. somewhat secund, often falcate, subulate-
setaceous, dull green, sometimes slightly toothed at
apex, nerve predominant; per. l. large, sheathing.
Caps. ovate, sometimes gibbous, 8-furrowed, lid red,
obliquely beaked; barren fl. gemmiform. Monoicous.

Fissures of alpine rocks. July.
Scotland, N. Wales, Yorkshire, Lakes, &c.

63. **D. schisti,** *Gunn (D. Blyttii,* ed. 1). St. branched, fastigiate. L. flexuoso-patent, or subsecund, from an erect base lanceolate-subulate, soft, entire, nerve excurrent into a fine point; per. l. sheathing, suddenly subulate. Caps. subcernuous, ovate, incurved, without striæ, strumose, lid obliquely rostrate.

Alpine and subalpine rocks. Not common. August. Scotland, N. Wales, Yorkshire.

64. **D. falcatum,** *Hedw.* St. shorter, dichotomously branched and fastigiate. L. strongly falcato-secund, from a lanceolate base, subulato-setaceous, concave, denticulate at apex, nerve excurrent. Caps. cernuous, shortly obovate, strumose, almost smooth when dry, lid large, beaked.

Alpine rocks.   August, September. Scotland, N. Wales, Yorkshire.

65. **D. Starkii,** *Web. and M.* St. 1—3 in., branched. L. subulate-setaceous from a lanceolate base, falcato-secund, entire, not crisped, nerve narrow, excurrent. Caps. oblong, cylindric, arcuate, gibbous, strumose, striate, subcernuous, with a long oblique beak.

Alpine rocks.   August. Scotland, Wales.

66. **D. molle,** *Wils. (D. glaciale,* ed. 1). St. 2—5 inches. L. erecto-patent, oblong-lanceolate, subulate, erecto-patent, purplish, margin incurved, nerve narrow, vanishing in apex. Caps. oblong-cylindric incurved, not striate, beak short, oblique.

Summits of highest Scotch mountains.

August, September.

Section 2. *EU-DICRANUM.*

Plants robust, tomentose, generally dioicous. L.

lanceolate, the longitudinal walls of their cells communicating by fine pores.   Caps. cernuous, cylindraceous, more or less arcuate.

a. L. not undulate.

67. **D. majus,** *Smith.*   St. 2—6 inches, loosely cæspitose.   L. lanceolate-subulate, falcato-secund, concave, dentato-serrate at apex and back, nerve excurrent. Caps. horizontally cernuous, curved, furrowed when dry, lid with a very long oblique beak, setæ pale, aggregate.

Shady banks, &c., in woods.   Frequent.

July, August.

68. **D. scoparium,** *Hedw.*   St. 2—4 inches, loosely tufted, dichotomous.   L. lanceolate-subulate, secund or falcato-secund, carinato-concave, margins inflexed, serrate at apex; nerve with about four prominent ridges at back, serrate at apex, upper cells linear-rectangular; per. l. larger, convolute.   Caps. cylindrical, slightly curved, seta solitary, lid with a long beak.

Shady banks and rocks.   Common.   July, August.

Var. β. ALPESTRE.   Denser.   L. denser, broader, slightly secund, margin and nerve scarcely toothed.

Killarney.

Var. γ. RECURVATUM.   Slender, elongate.   L. suddenly larger above, falcato-secund.

Sussex.

Var. δ. TURFOSUM.   Tall, glossy, with few radicles. L. erecto-patent, subcuspidate at apex, almost entire, dark-coloured below.

Lofthouse, Yorkshire.

Var. ε. ORTHOPHYLLUM.   Tufts dense, radiculose.   L. nearly erect, rigid, elongate, almost entire at apex.

Var. ζ. PALUDOSUM. Dense bright green tufts, much radiculose. L. short, broad, subsecund, sharply serrate and rugulose at apex.

Scotland, Derbyshire.

b. L. transversely undulate.

69. **D. Bonjeanii,** *De Not.* (*D. palustre,* ed. 1). St. 3—6 inches, erect, branched, subfastigiate. L. spreading, subsecund, linear-lanceolate, acute, terminal ones crowded into a cuspidate cluster on the barren shoots, serrate at apex, nerve thin and narrow, not reaching to apex, smooth at back or faintly serrate near apex, their upper cells elongate. Caps. suberect, slightly curved, subcylindrical, striate.

Marshy places and moist banks. Aug., September.

Var. β. JUNIPERIFOLIUM. With shorter, wider, and more rigid leaves.

Var. γ. CALCAREUM. St. shorter, more rigid. L. secund, subfalcate, margins incurved, slightly undulate and serrate only at apex.

Rare.

Sussex.

70. **D. Bergeri,** *Bland.* (*D. Schraderi,* ed. 1). St. 3—6 inches. L. broadly lanceolate, subsecund, rather obtuse, carinato-concave, subrugose, toothed on margin and keel, subpapillose at back near apex, upper cells small, quadrate, nerve vanishing at apex. Caps. oval-oblong, incurved, lid rostrate.

Turfy bogs. Rare. August, September.

Risley Moss and Wybunbury Bog, Cheshire.

71. **D. spurium,** *Hedw.* St. 2—4 inches, loosely cæspitose, densely tomentose. L. ovate-lanceolate, acuminate, eroso-denticulate, papillose at back, not

nerved to apex. Caps. subcylindrical, arcuate, slightly strumose, striate, lid with a long curved beak.

Moors and bogs. Not common. July.

Section 3. *APORODICTYON.*

Plants of medium size, radiculose. L. lanceolate-subulate, their cell walls not interrupted by pores. Caps. cernuous or erect, cylindraceous, curved or symmetric.

a. Caps. cernuous, curved.

72. **D. congestum**, *Brid.* St. erect, 1—2 inches, densely tomentose. L. crowded, linear-lanceolate, secund, coarsely serrate above, nerve narrow, thin, scarcely reaching apex, where it is sometimes distantly toothed, upper cells larger than in *fuscescens*, and irregular in form. Caps. ovate-oblong, with a long-beaked lid. Dioicous.

Mountain rocks. Very rare. August.
Ben Lawers, 1878 (Boswell).

Var. β. FLEXICAULE. St. longer, flexuose. L. longer, almost entire.

Rare.

Ben Lawers, Lochnagar, Teesdale.

73. **D. fuscescens**, *Turn.* St. 2—4 inches, loosely tufted, tomentose. L. spreading, subsecund, flexuose, canaliculate, minutely toothed at apex, nerve broad, excurrent, forming the whole of the subula, lower cells rectangular, upper small, quadrate, papillose at back. Caps. oblong, incurved, furrowed when dry, lid with a very long beak. Dioicous.

Alpine and subalpine rocks. Frequent. August.
Var. β. FALCIFOLIUM, *Braith.* "L. all falcato-secund,

flexuoso-cirrhate towards apex, shorter and less attenuated to point."

Teesdale, Dunoon.

74. **D. elongatum,** *Schleich.* St. 3—6 inches, tomentose. L. oblong-lanceolate, subulate, incurved above, margin entire or nearly so, lower cells rectangular, elongate, upper smaller, oblong. Caps. ovate, gibbous, scarcely striate, lid with a long oblique beak.

Peat in mountainous places. Rare. August.
Braemar and Inverness-shire.

b. Caps. erect, symmetric.

75. **D. montanum,** *Hedw.* (*Weissia truncicola,* ed. 1). In large, dense, bright green tufts. St. 1—2 inches, dichotomous, reddish, radiculose below. L. erect when moist, and often secund on the young shoots, rather soft, papillose at back, from a narrowly lanceolate base, gradually subulate, channelled, thinly nerved nearly to apex, margin not revolute, sharply denticulate above and on the back of the nerve, strongly cirrhate and twisted when dry, basal cells large, cylindraceo-vesicular, the rest small, quadrate.

Trunks of trees and roots in woods. Rare.

Several woods in Midland counties; Den of Airlie; Wharncliffe woods (?) Dr. Parsons.

76. **D. flagellare,** *Hedw.* St. ½—1 inch, branched, tomentose, giving off flagellæ, whose leaves are small, upright, obtuse. L. lanceolate, secund, spreading, tufted at summit of stem, curled when dry, elongated into a smooth, almost tubular subula, slightly toothed at extreme apex, where nerve vanishes. Caps. reddish, striate.

Trunks and roots of trees. Very rare. August.

Kent (Holmes), 1874.

N.B.—Nos. 75, 76 are hitherto sterile in Britain.

77. **D. viride,** *Sull. and Lesq.*   Dioicous, in dense
cushions, or cæspitose, reddish and tomentose at base,
above dark green.   Branches dichotomous.   L. oblong-
lanceolate, subulate, scarcely twisted when dry, basal
cells enlarged, lax, hyaline, above small, quadrate, not
serrulate, nerve running out in the concave awl; per. l.
sheathing.   Caps. erect, oblong, slightly incurved, lid
with a long beak.

Trunks of trees, &c.   Rare.   July, August.
Staffordshire (Mr. Bloxam), 1864.

78. **D. Scottianum,** *Turn.* (*D. Scottii,* Br. M. Fl.).
St. 2 or 3 inches, robust.   L. erecto-patent, subsecund,
incurved, lanceolate-subulate, slightly twisted at apex
only when dry, concave, entire, basal cells elongate,
thickened, upper small, quadrate, nerve strong, excurrent.   Caps. elongate, slightly curved, tapering at
base, lid obliquely rostrate.

Rocks in mountainous districts.   Not common.

July, August.
Ireland, South of England, Argyle, &c.

79. [**D. Sauteri,** *B. and S.*   Differs from above in its
leaves being more subulate, slightly serrulate at apex,
and the basal alar cells very long.

Its var. β. CURVULUM, which has its caps. horizontal
and curved, is supposed to be British, from specimens
in Herb. A. O. Black (Braithwaite).]

80. **D. longifolium,** *Hedw.*   Cæspitose, tufts pale
green or whitish.   St. arcuate or geniculate, ascending,
slightly radiculose.   L. long, falcato-secund, rarely
spreading, longly subulate from a lanceolate base,
nerve very broad, margin and back serrate at apex;

per. l. convolute, sheathing. Caps. elongate-cylindrical, upright or subincurved, without striæ, brown, beak subulate. Dioicous.

Subalpine rocks. Rare. Autumn.

Ben Lawers, 1866 (Dr. Stirton), Dumfries, Glen Prosen.

81. **D. asperulum,** *Mitt.* (*Dicranodontium aristatum,* Auct.). Dioicous. Tufts 2—3 inches, tomentose. L. falcato-secund, subulate from an ovate base, nerve occupying the whole of the long subula, serrulate on margin and spinulose at back, cells lax, pellucid, larger and hyaline at basal angles. Caps. on a long, twisted, yellow seta, oval-cylindric, plicate when old, lid with a long, straight, subulate beak.

Mountainous sandstone rocks.

Scottish mountains.

82. **D. uncinatum,** *Harv.* (*D. circinnatum,* ed. 1). Dioicous, in loose, irregular, light green tufts. St. 3—6 inches, dichotomous, geniculate or ascending, radiculose. L. very long, secund, arcuate from an oblong sheathing base, decurrent at angles, longly subulate, concave, nerve flattened, covering one-fifth of base and all the subula, which is denticulate at back and margins, base laxly areolate in middle, with narrower cells at margin.

Fr. unknown.

Ben Voirlich, Clova, Ben Nevis, &c.

18. **DICRANODONTIUM,** *Br. and S.**

83. **D. longirostre,** *B. and S.* (*Didymodon denudatus,*

* I have felt myself obliged to use this generic name for this species, although Dr. Braithwaite, rightly I think, prefers the older name *Didymodon* for it, but as that is used further on for other species, I cannot adopt it here.

Br. M. Fl.). St. 1—3 inches, blackish. L. deciduous, falcato-secund, subulato-setaceous, from an ovate, sheathing, auriculate base, denticulate above on predominant nerve. Caps. elliptic-oblong, smooth, on a thick curved or flexuose seta, with a long straight beak.

Mountainous woods. Rare.   October.

Scotland, N. Wales, Ireland, Cheshire, Yorkshire, Lancashire.

Var. β. ALPINUM, *Schp.* (*Campylopus alpinus*, ed. 1). Taller and more robust. L. not deciduous, erect or subsecund, rather rigid (Braithwaite).

Scotland, Yorkshire, N. Wales, Ireland.

19. CAMPYLOPUS, *Brid.*

Section 1. *Leaves without hyaline points.*

a. L. not auricled at base.

84. **C. pyriformis**, *Brid.* (*C. turfaceus*, B. and S.). Tufts flat, olivaceous or bright green, finally tawny. St. ½—1 inch, slender, erect, radiculose only at base. L. less crowded, gradually larger upwards, erecto-patent, lower lanceolate, middle lanceolate-subulate, upper from a lanceolate base, setaceous, nerve one-third base, thin, channelled at back. Areolæ resembling *C. flexuosus*, but thinner, hyaline at base. Fr. several from same apex. Caps. ovate, olivaceous, fulvous when ripe, sulcate, lid obliquely rostrate. Calyptra whitish, tip brown.

Moist heaths and sides of ditches.   Winter, spring.

85. **C. fragilis**, *B. and S.* Tufts pale green, glossy. St. ½—2 inches, fragile. L. densely crowded, erecto-patent, rigid, incumbent when dry, lower lanceolate, upper extended into a subula, toothed at apex, wings

recurved above, nerve very broad. Basal areolæ lax, pellucid, narrow, rectangular, above minute, quadrate, no distinct alar cells. Caps. solitary, bent down, oval, symmetric, fuscous, when dry plicate, contracted below the mouth, lid conico-subulate, oblique, red. Calyptra whitish, rufous at apex.

Sandstone rocks and moist heaths.

Var. β. DENSUS, *B. and S.* St. taller. L. shorter, with more acute, entire points and laxer cells.

86. **C. Schimperi,** *Milde.* Tufts dense, compact. St. 1—2 inches, slender, light silky green above, fuscous below. L. erecto-patent, appressed when dry, straight, rigid, lanceolate-subulate, channelled, denticulate only at apex, nerve very broad, basal cells lax, rectangular, hyaline, very narrow at margin, above elliptic.

Alpine hills.

Scotland.

87. **C. subulatus,** *Schpr.* (*C. brevifolius,* ed. 1). St. ½ inch, yellowish-green, not radiculose, with caducous ramuli. L. short, rigid, erect, lanceolate, longly acuminate, concave, obsoletely toothed at apex, nerve half base, basal areolæ hyaline, lax, rectangular, gradually shorter and more quadrate, lower ones with their transverse walls much thickened.

Dry and stony places.

Scotland, Ireland.

Var. β. ELONGATUS, *Boswell.* " Tufts broad, extensive, solid, dense. St. slender, elongate, 1—2 inches, copiously radiculose below, repeatedly innovating, with fasciculate branches above. Br. without radicles. L. as in type.

" Muddy banks of R. Wye, near Builth."

H. Boswell, " Naturalist," ix. p. 28 (Sept., 1883).

b. L. auricled at base.

88. **C. Swartzii,** *Schpr.* Tufts dense, soft, yellowish-green, brownish below, without radicles. St. 2—3 inches, slender. L. erecto-patent, straight or slightly secund, lowest lanceolate, upper lanceolate-subulate, entire at apex, base somewhat sheathing, auricles hyaline, inflated, decurrent, nerve two-thirds of base, finely sulcate at back towards apex, basal areolæ narrow, auricular, very lax, hexagono-rectangular, hyaline, above subquadrate.

Granite alpine rocks.

Wales, Scotland, Ireland.

89. **C. Shawii,** *Wils. MS.* Tufts lax, yellow-green above, blackish-brown below, 1—2 inches high. St. robust, with numerous radicles. L. erecto-patent, straight, rigid, from a somewhat contracted, linear base, lanceolate, longly subulate, suddenly narrowed at one-third their length, margin involute above, apex acute, with a few minute denticulations, nerve two-thirds width of base, cells at basal wings enlarged, lax, reddish-brown, exterior rows hyaline, above rectangular, and then rhomboido-elliptic. L. falcate when growing in dry places.

Outer Hebrides, 1866 (Mr. Shaw).

· 90. **C. setifolius,** *Wils.* Tufts lax, soft, bright or yellowish-green above, blackish below, without radicles. St. 5—10 inches, slender, erect, geniculate. L. distant, erecto-patent or subsecund, glossy, from a lanceolate base, gradually running into a very long subula, sometimes half twisted, uppermost with wings serrate, nerve half width of base, auricles very large and inflated, the cells partly fuscous, partly hyaline, hexagonal, above hexagono-rectangular, upper rhombic, chlorophyllose. Flowers of each sex collected in capitula, males 3—4, females numerous.

Wet places, and clefts of rocks.

Scotland, Ireland.

91. **C. flexuosus,** *Brid.* Tufts dense, yellowish-green. St. ½—1½ inches high, erect, dichotomous, with rufous purple radicles to apex, bearing gemmæ intermixed. L. patent, straight, or secund, subfalcate, lower lanceolate, upper subulate, uppermost very long and toothed at apex, all concave, glossy, red when old, nerve one-third width of base, angles not decurrent, with short, wide, fuscous cells, others hexagono-rectangular, upper quadrate and' chlorophyllose ; per. l. sheathing longly subulate, with a narrower nerve. Calyptra fuscous at apex. Caps. oval, regular, or gibbous, short-necked, olivaceous, with eight striæ, sulcate when dry, lid conico-rostrate. Annulus broad, double.

Subalpine moist rocks and peaty soil. November.

92. **C. paradoxus,** *Wils. MS.* Tufts ½—1 inch high, fastigiate, dull yellowish-green above, pale brown below. St. with short lateral ramuli, and few rufous radicles. L. erecto-patent (erecto-appressed when dry), uppermost longest, slightly secund, lanceolate-subulate, concave, apex usually of two teeth, with a few irregular ones below on each side, nerve one-third width of base. Lamina extended to apex, basal cells thin, enlarged, hyaline when young, afterwards fuscous, above rectangular, in 14—16 longitudinal rows, thickened and quadrate towards apex.

Peaty soil.

Cheviots (barren), Boyd and Hardy, 1868; Rombold's Moor, Yorkshire; Arran, 1883 (Rev. A. Ley).

Section 2.　*Leaves with hyaline points.*

93. **C. atrovirens**, *De Not.* (*C. longipilus*, Brid., pro
parte; Wils., Bry. Brit.; et Schimp., Musc. Eur. Nov.).
Dense tufts 1—3 inches high, above yellowish-green,
below brownish, at base black.　St. erect, dichotomous,
with few radicles at base.　L. lower, lax, shorter, the
rest densely crowded, erecto-patent, lanceolate, very
longly subulato-setaceous, channelled below, auricled,
nerve excurrent into a hoary hispid arista, channelled
at back, one-third width of leaf base, cells of auricles
dilated, castaneous, central colourless, above these
subrectangular, uppermost oblongo-elliptic.　Female
fl. 2, 3 at apex of innovations.

Wet rocks, and moorlands in mountainous districts.

Var. *β*. FALCATUS, *Braith.*　" St. short, more robust.
L. dense, broader, falcato-secund, circinnate, very
concave."

Connemara (Barker); Islay, September, 1883 (Rev.
A. Ley).

94. **C. introflexus**, *Brid.* (*C. longipilus*, Bry. Eur. pro
parte; *C. polytrichoïdes*, De Not.; *D. ericetorum*, Mitt.).
Densely tufted, olivaceous brown below, innov. yellow-
green with hoary tips, $\frac{3}{4}$—$1\frac{1}{2}$ inches high, sparingly
radiculose, dichotomous.　L. imbricated, erecto·patent,
lanceolate-subulate, channelled, not auricled, wings
but little incurved, comal leaves broader, lanceolate-
acuminate, lowest muticous, rest prolonged into a
diaphanous spinuloso-denticulate arista shorter than
the leaf, nerve three-fourths width of limb, lammelluli-
gerous at back, basal cells hyaline, large, and empty,
gradually becoming obliquely oval and minute, chlorophyllose, a few fuscous alar cells in comal leaves;

per. l. oblong, convolute, subulate at apex. Thecæ aggregated on short peduncles, oval, unequal, rough at base, lid obliquely rostrate. Calyptra reaching middle of capsule, sparingly fimbriate.

Dry heaths and stony places.

Cornwall, Jersey, Scotland, Ireland.

95. **C. brevipilus,** *B. and S.* In dense, broad tufts, when dry glossy yellow-green above, fuscescent below, $\frac{3}{4}$—$1\frac{1}{4}$ inches high, almost free from radicles, fastigiate. L. erect, densely crowded, narrowly lanceolate-subulate, very concave, the point denticulate at margin and back, scarcely auricled, nerve one-third width of leaf, base excurrent into a short hair point; per. l. wider, sheathing, narrowed into a hispid hair, margin recurved above base. Areolæ lax, basal cells quadrate, above rhomboidal, flexuose, marginal very narrow. Female fl. solitary.

Heathy places.

Sussex, Hants, Cheshire, York, Arran.

Var. AURICULATUS, *Ferg.,* Rev. Bry., 1879, p. 26. Having more or less conspicuous auricles, composed of large, fuscous cells.

Several places in Scotland and England.

Tribe vi. LEUCOBRYACEÆ.

20. **LEUCOBRYUM,** *Hampe.*

96. **L. glaucum,** *Hampe.* St. 1—6 inches or more, dichotomous, fragile, fastigiate. L. subulate from an ovate-lanceolate base, erect, rather obtuse, and apiculate. Caps. cernuous, strumose, furrowed when dry.

Moist heaths, woods, rare in fr.   Winter.

Tribe vii.   FISSIDENTACEÆ.

21. FISSIDENS, *Hedw.*

a. Fruit terminal.

97. **F. exilis,** *Hedw.* St. ⅛ inch, simple. L. few, lower small, ovate, upper lanceolate-oblong, oblique, acute, margin not bordered, serrulate, nerved to apex, dorsal wing not reaching to base of leaf. Caps. elliptic-oblong, erect, lid conical, obliquely rostrate. Monoicous.

Shady banks and woods. Not frequent.

E. S. I.   January—March.

98. **F. pusillus,** *Wils.* St. shorter. L. erect, acute, narrowly lanceolate, nerved to apex, border narrow. Caps. suberect, ovate, cylindric, lid with an oblique beak. Dioicous.

Sandstone rocks.   August, September.

Var. *β.* LYLEI. L. scarcely margined, broader. Rare.

Cheshire, Warwick, Hereford, Scotland.

Var. *γ.* MADIDUS. L. longer and narrower. Caps. subcylindric.

Castle Howard, Spruce, 1844.

99. **F. incurvus,** *Schw.* St. ¼ inch about, ascending from a decumbent base. L. lanceolate, oblong, apiculate, narrowly margined, nerve ceasing near the serrate apex. Caps. oval, oblique, curved, cernuous, lid conical, rostellate. Perist. not immersed. Barren fl. sessile at base of stem.

Shady banks.   February, March.

Cheshire, Yorkshire, Snssex, Oxon, Dorset, &c.

Var. *β.* TAMARINDIFOLIUS, *Donn.* St. slender. L. elliptical, "subfalciform apiculate," with an entire, pellucid, cartilaginous border, nerved to apex. Caps. ovate-oblong, curved, inclined, lid short, conical, with a bluntish point.

100. **F. viridulus,** *L.* St. ¼ inch about. L. lanceolate, acute, entire, bordered, crisped when dry, dorsal wing not reaching to base, nerved nearly to scarcely denticulate apex. Caps. oval-oblong, erect, lid conical, with a blunt point. Perist. immersed. Barren fl. on a short branch, at base of fertile stem.

Shady banks, rivulets on stones, &c. Winter.

Var. β. FONTANUS (*F. crassipes*, ed. 1). St. ¼—¾ inch. Plant more robust. L. larger, broader, and more numerous, scarcely nerved to apex.

Sluices. October, November.

101. **F. bryoides,** *Hedw.* St. ¼—½ inch. L. lanceolate, acute, apiculate, with a thickened margin, dorsal wing reaching to and broad at the base, strongly nerved to or beyond apex. Caps. elliptical, erect, symmetrical, lid conical, acutely rostellate. Barren fl. axillary.

Shady banks. Frequent. Winter.

Var. β. CÆSPITANS. Taller. L. with a narrower margin, apex slightly denticulate. Caps. ovate.

Penzance and Kymal Bridge.

102. ["**F. Orrii,** *Lindb.* Very small. L. narrow, linear, very acute, with a thickened border and excurrent nerve. Caps. minute, obovate, cernuous, lid conico-rostrate." Br. M. Fl., 73.]

Near Glasnevin Botanic Gardens. Probably alien.

103. **F. Osmundioides,** *Hedw.* St. 1—2 inches, tufted, erect, radiculose. L. lower scattered, small, upper larger, crowded, ovate-lanceolate, obtuse, apiculate, margin not bordered, almost entire, not nerved to apex,

the latter sometimes toothed. Caps. small, oval-oblong, suberect, lid large, convex, rostrate. Dioicous.

Wet mountainous rocks. July.

104. ·F. rufulus, *Br. and Sch.* St. ½—1½ inches, dichotomous, radiculose. L. crowded, erecto-patent, cultriform, vertical lamina prolonged to base, entire, strongly nerved to apex, which is eroso-denticulate, nerve and broad border orange-red. Caps. erect, ovoid, lid conical, obtuse.

Rocks and stones in streams.

Westmoreland.

105. F. serrulatus, *Brid.* St. 1—3 inches. L. crowded, straight in outline and obtuse, pointed, margin of conduplicate portion finely serrulate, also apex of lamina, border of large, yellowish cells, thickly nerved nearly to apex. Caps. oval, inclined, lid with a long, straight beak, seta thick, short, flexuose.

Damp shady places. Winter.

Near Penzance (Curnow).

b. Fruit lateral.

106. F. decipiens, *De Not.* St. about ½ inch, fasciculate from base. L. lower distant, coulter-shaped, upper imbricate, patulous, oblong-ligulate, acute, or mucronulate, dorsal wing narrow, nerve strongly excurrent, excavate, the cultriform lobe of the upper leaves obliquely acute, longer than half the leaf, upper part strongly serrate. Caps. ovate, somewhat constricted at base, erect or inclined, lid large, rounded, beaked. Dioicous. [De Notaris, Epilogo Briol. Ital., 1869, p. 480.]

Damp rocks and old walls. Winter.

Oxford, Cornwall, Westmoreland, Wales, Scotland Ireland.

107. **F. taxifolius,** *Hedw.* St. about ½ inch, fasciculate from base. L. lanceolate, pointed, not bordered, finely crenulate, nerved almost to apex. Caps. almost ovate, inclined on a seta curved at summit, and inserted at base of stem, lid large, convex, with a long, oblique beak. Monoicous.

Moist shady banks. December—February.

108. **F. adiantoides,** *Hedw.* St. 1—3 inches, branched, leafy. L. ovate-lanceolate, finely serrulate below, dentate at apex, nerved almost or quite to apex, border sometimes thickened. Caps. oval-oblong, constricted at mouth when dry, cernuous, arising from middle of stem, lid with a long beak. Monoicous.

Shady wet rocks and bogs. October—April.

109. **F. polyphyllus,** *Wils.* St. 3—3 inches, simple or branched. Br. arcuate. L. elongate, lanceolate, rather acute, strongly nerved to serrulate apex, not bordered. Caps. cylindric, inclined, tapering at base. Barren fl. numerous, axillary.

Moist shady rocks on mountains.

Wales, Ireland, Cornwall.

Tribe viii. SELIGERIACEÆ (Bruchiaceæ).

Fam. 1. Seligeriæ.

22. **SELIGERIA,** *B. and S.*

a. Peristome absent.

110. **S. Doniana,** *Muell.* (*Anodus,* ed. 1). St. minute, ⅛ inch, gregarious. L. almost setaceous, lanceolate-subulate, very minutely toothed at base; per l. bluntish and rather shorter. Caps. cup-shaped or turbinate, mouth wide. Calyptra dimidiate. Perist. none, lid with a short beak.

Sandstone rocks. Rare. September.

b. Peristome present.

111. **S. pusilla,** *Bruch. and S.* Minute, ⅛ inch stems, loosely tufted, simple or dichotomous. L. lanceolate-subulate, setaceous above, very narrow, thinly nerved nearly to apex. Perist. with teeth distantly barred. Caps. on an upright pedicel, turbinate when dry, with a flattish, oblique, beaked lid.

Shady limestone rocks.   April, May.

112. [**S. acutifolia,** *Lind.* Very small. L. and per. l. from a more or less sheathing base, abruptly narrowed into a subterete, setiform, acute, pointed awl, formed by the excurrent nerve, crenulate, seta 1 mm. long. Caps. small, scarcely exserted, pyriform, with a short neck, lid with a short, scarcely oblique beak] ;—type not British but

Var. *β.* LONGISETA, *Lindb.* Plant larger, seta 2—3 mm. long, caps. exserted, beak of lid longer and more oblique—gathered by Mr. Wilson, 14th May, 1831, and sent by him to Dr. Lindberg.

Derbyshire, Yorkshire.   May, June.

113. **S. trifaria,** *Brid.* (*tristicha*, ed. 1, and Schp. Syn.). Densely cæspitose, rigid. L. exactly trifarious, crowded, rigid, shortly lanceolate, muticous, base whitish. Caps. yellowish-brown, subspherical, with a tumid neck, lid large, with a long, oblique or arcuate beak ; per. teeth narrower than in *calcarea.*

Calcareous stones and rocks.   Summer.

Blair Athol (Miss McInroy), Glen Tilt, and Ben-y-Gloe ; Yorkshire, Derbyshire.

114. **S. paucifolia,** *Dicks.* (*S. subcernua*, Schp. ; *S. calcicola*, Mitt.). Densely gregarious, low. L. crowded, erecto-patent, lower ones lanceolate, upper subulate, from a narrow, oblong base, margins plane, nerve

exserted, areolæ dense, rectangular. Caps. elliptical, subcernuous, on a long seta, unsymmetrical, lid with a long beak. Male fl. at base of female plant.

Limestone rocks and stones. June.

Chalk Downs, Sussex, Mr. Mitten ; near Wetherby, 1801 (Dickson); Surrey; Kent.

115. **S. calcarea**, *B. and S.* St. short, more robust than No. 111. L. ovate-subulate, obtuse, dull green, with a thicker nerve. Caps. turbinate, shortly beaked, on a short, stiff seta. Perist. teeth broader, obtuse, closely barred.

Chalk cliffs. April, May.

Kent, Sussex, Surrey, Beds.

116. **S. recurvata**, *B. and S.* St. minute, gregarious. L. lanceolate-subulate, somewhat flexuose, acute, nerve excurrent generally. Caps. obovate, elliptical, seta curved, drooping.

Sandstone rocks. Rare. April, May.

Fam. 2. Brachydontæ.

23. CAMPYLOSTELIUM, *Bruch. and S.*

117. **C. saxicola**, *B. and S.* Minute. L. elongate, linear-lanceolate, crowded, entire, twisted, nerved nearly to summit. Caps. elliptical, drooping, on a geniculate pedicel, annulus double. Calyptra 5-cleft at base.

Sandstone rocks. Rare. November.

24. BRACHYDONTIUM, *Bruch.* (*Brachyodus*, Nees, ed. 1).

118. **B. trichodes**, *N. and H.* Very minute. L. lanceolate-subulate, almost setaceous, erect, with an excurrent nerve forming half the leaf. Caps. cylindric,

erect, furrowed. Perist. very short, annulus large, lid flattish, with a long, straight beak.

Subalpine sandstone rocks. Spring.

Fam. 3. Blindiæ.

25. **BLINDIA,** *B. and S.*

119. **B. cæspiticia,** *Schwaeg.* (*Stylostegium,* ed. 1). St. ¼—½ inch, densely tufted. Br. fastigiate. L. somewhat falcate and secund, ovate-lanceolate, acuminate; per. l. larger, with a sheathing base, entire, nerve predominant. Caps. immersed, roundish-pyriform, glossy. Perist. absent, lid obliquely beaked, adherent to columella.

Alpine rocks, in crevices. Rare. July—September.

120. **B. acuta,** *Huds.* St. ½—3 inches, tufted. L. subulate or lanceolate-setaceous, rigid, glossy, subsecund, entire, nerve thin; per. l. sheathing. Caps. roundish-pyriform, on a short, reddish pedicel. Perist. simple, 16 entire or perforate teeth, lid with a longish, oblique beak.

Moist alpine or subalpine rocks. Summer.

Tribe ix. LEPTOTRICHACEÆ.

Fam. 1. Bruchiæ.

a. Nerve narrow. L. cells large, lax.

26. **PLEURIDIUM,** *Brid.*

121. **P. nitidum,** *Hedw.* (*P. axillare,* Br. M. Fl.). L. generally erect, linear-lanceolate, keeled, subdenticulate near apex, nerved nearly to summit. Caps. elliptic ovate, with a short, oblique point, sometimes pendulous, on a short pedicel.

Moist banks, &c. Winter.

Var. β. STRICTUM. Smaller. L. narrower, straight. Caps. nearly spherical.

Scotland.

b. Nerve broad, cells small.

122. **P. subulatum,** *Huds.* St. ⅛ inch. L. lanceolate, sharply tapering from a broadish base, not keeled, serrulate nerve ceasing near the apex; per. l. almost setaceous. Caps. roundish-ovoid, pale brown, immersed, on a very short pedicel.

Banks and fields. Common. Spring.

123. **P. alternifolium,** *Kaulf.* St. sometimes with innovations, ½ inch long, or more; st. l. lanceolate acuminate from a broad base; per. l. subulate-setaceous, nerve excurrent, and forming nearly all the upper half of the leaf. Caps. ovoid, immersed, brownish, with an oblique point.

Banks and fallow ground. Spring.

Fam. 2. **Ditricheæ.**

27. **DITRICHUM,** *Timm.*

a. L. squarrose. Caps. cylindric.

124. **D. tenuifolium,** *Schrad.* (*Ceratodon cylindricus,* ed. 1). St. ¼ inch, gregarious. L. subulate from a dilated, ovate, amplexicaul base, flexuose, minutely toothed above, nerve forming the whole subula. Caps. cylindrical, smooth, erect or slightly curved, on a pale, slender seta, lid conical.

Sandy banks. Rare. May, June.

Yorkshire, Cheshire, Lancashire, Sussex, Scotland, Ireland.

125. **D. tortile,** *Schrad.* St. ¼ inch, gregarious, subflexuose. L. mostly secund, somewhat falcate, lanceo-

late-subulate, margin reflexed, nerve excurrent into the
slightly toothed apex.   Caps. small, cylindrical, erect,
regular or curved, lid conical, slightly rostrate.   Perist.
teeth irregular, purplish-red.   Dioicous.

Sandy places.   Rare.   October, November.
Yorkshire, Sussex.

Var. β. PUSILLUM.   Shorter and denser.   L. short,
nearly straight.   Caps. oval.

Near Belfast ; Castle Howard.

126. **D. homomallum,** *B. and S.*   St. scarcely ¼ inch,
cæspitose.   L. subulato-setaceous from a broadish base,
mostly secund, nerve broad, much excurrent, entire.
Caps. erect, oblong-ovate, brown, on a long red seta.
Annulus present, lid short, conical, obtuse.   Basilar
membrane of perist. very narrow.   Dioicous.

Sandy banks.   Autumn.

Var. β. ZONATUM.   St. elongate, ½—2 inches.   L.
shorter, scarcely secund, erecto-patent.   Caps. smaller.
Tufts green above, brownish below.

Scotch and Welsh mountains.

127. **D. subulatum,** *Bruch.*   St. ¼ inch, cæspitose.
L. subulato-setaceous from an ovate base, spreading or
somewhat secund, entire, with a long excurrent nerve.
Caps. oval, erect, lid large, obliquely rostellate, seta
somewhat flexuose.   Annulus none.   Basilar membrane
very narrow.   Antheridia axillary, naked.

Cornwall (Rev. — Tozer), R. V. Tellam.   Spring.

128. **D. flexicaule,** *Schleich.*   St. 1—3 inches,
flexuose, cæspitose, with fastigiate branches.   L. very
long and setaceous, flexuose, concave, usually secund,
nerve broad, excurrent, toothed at apex.   Caps. erect,
small, ovate-oblong.   Annulus present.   Perist. teeth
long, irregular.   Dioicous.

Scotch and Derbyshire mountains (calcareous). June.
Var. *β*. DENSUM. Densely cæspitose. L. erect, shorter.

129. **D. glaucescens,** *Hedw.* St. ½ inch. Branches fastigiate. L. glaucous, linear-lanceolate, margin plane (upper crowded into a tuft or coma), nerve sometimes excurrent into the denticulate apex. Caps. oblong-oval, pale brown, with a long, beaked lid. Basilar membrane very narrow. Barren fl. gemmiform.

Scotch mountains. Summer.

Fam. 3. Ceratodontæ.

28. CERATODON, *Brid.*

130. **C. purpureus,** *Brid.* St. ¼—2 inches, cæspitose, branched. L. oblong-lanceolate, margin recurved, nerve excurrent. Caps. elliptic-oblong, irregular, purple, angular when dry on a purplish-red seta, lid conical.

Banks, &c. Common. April, May.

Fam. 4. Distichiæ.

29. DISTICHIUM, *B. and S.*

131. **D. capillaceum,** *B. and S.* (*Swartzia montana,* Br. M. Fl.). St. 1—2 inches, cæspitose, shining green above, ferruginous below. L. subulate-setaceous, spreading, integrate, nerve excurrent. Caps. erect, ovate-oblong or almost cylindrical, reddish-brown. Perist. teeth narrow, articulate, bi- or trifid, lid conical.

Scotch and Welsh mountains. Summer.

132. **D. inclinatum.** *B. and S.* (*S. inclinata,* Br. M. Fl.). St. shorter than last, and less cæspitose. L. narrower and minutely serrate at apex ; per. l. 1, 2, or

3 together.   Caps. oval, olive-brown, inclined or cernuous.   Perist. teeth larger, lanceolate, articulate,
entire or perforate, bi- trifid.

Irish and Scotch mountains.   June, July.

Tribe x.   POTTIACEÆ.

Fam. 1.   **Phasceæ.**

30.   **EPHEMERELLA,** *Muell.*

133. **E. recurvifolia,** *Dicks* (*Phascum,* ed. 1).   Minute.
St. almost none.   L. lingulate, rarely linear-lanceolate,
erect, frequently recurved, denticulate at the apex,
with a strong, generally excurrent nerve, lowest ovate-
lanceolate, nerveless.   Caps. roundish-ovate, immersed,
lid with a short, oblique beak.

Heaths and fallows.   Autumn, winter.

31.   **MICROBRYUM,** *Schpr.*

134. **M. Floerkeanum,** *Web. and M.* (*Phascum,* ed. 1).
Almost stemless, very minute.   L. erecto-patent,
broadly ovate, tapering to a point, lower ones small,
nerveless, upper ones larger, nerve excurrent, margins
reflexed.   Caps. ovate-spherical, shortly beaked, immersed with subconical, curved-pointed calyptra.
Antheridia naked, axillary.

Clay or chalky fields.   Rare.

                    September—November.

32.   **SPHÆRANGIUM,** *Schpr.*

135. **S. muticum,** *Schreb.*   Minute, almost stemless.
L. lower ovate-acuminate, recurved, nerveless, middle
convolute, oblong, ovate, acuminate, recurved, concave,
nerve exserted, two or three uppermost concave, nerved

to an erose mucro.   Caps. round, reddish, erect, subsessile.

Moist banks and fallows.   Autumn, spring.

Var. β. MINUS.   Leaves entire.

Seaside.

136. **S. triquetrum,** *Spruce.*   Almost stemless. L.
in three rows, lowest minute, ovate, nerveless, medial
broadly ovate, nerved, three uppermost (perichætial)
cucullate, pointed, obovate, keeled, margins reflexed,
denticulate above, nerve excurrent in a recurved mucro.
Caps. spherical, horizontal or drooping, seta long,
slender, suddenly bent near its union with the capsule.

Cliffs, Sussex coast.   March.

33. PHASCUM, *Linn.*

a. M fl. axillary, gemmiform.

137. **P. cuspidatum,** *Schreb.*   From ⅛—¼ inch high.
St. simple or branched.   L. lower ovate-lanceolate,
upper oblong-lanceolate, all cuspidate, erect, concave,
keeled, with the nerve prominently excurrent, margin
recurved below, integrate.   Caps. roundish, immersed
on a short seta.

Moist banks, hedges, and fields. Common.   March.

Var. β. MACROPHYLLUM.   Leaves longer, lanceolate.
Caps. smaller.

Var. γ. SCHREBERIANUM.   St. elongated.   Br. dichotomous. L. distant, spreading.

Var. δ. PILIFERUM.   Pedicel curved.   L. with long
white filiform points.

Var. ε. CURVISETUM.   Caps. laterally exserted, subpendulous on a longish curved seta.   Upper leaves
lanceolate, cuspidate.

138. **P. bryoides,** *Dicks.*   St. ⅛—¼ inch, simple or

branched. L. lower ovate, pointed, upper elliptic, ovate, concave, erect, margin reflexed, pointed with the excurrent nerve. Caps. elliptical, with an oblique, blunt point, brown, exserted. Barren fl. sometimes terminal on a short branch.

Banks and fields. Rare. Spring.

Var. β. PILIFERUM. L. piliferous.

Var. γ. CURVISETUM. Seta curved, longer. Caps. cernuous.

Var. δ. BRACHYCARPUM. Caps. roundish, seta very short.

Var. ε. ATROVIRIDE. Smaller, with piliferous leaves.

Var. ζ THORNHILLII. "L. spreading, subreflexed, spathulato-lanceolate, margin plane, nerve slightly excurrent. Caps. narrowly elliptical, rostrate, pedicel elongated."

Near Newcastle.

b. Antheridia axillary, naked.

139. **P. curvicollum,** *Hedw.* St. short, reddish. L. erecto-patent, lower ovate acuminate, upper lanceolate, tapering, pointed with an excurrent nerve, entire, margin reflexed. Caps. roundish, blunt-pointed, cernuous, exserted, on a longish curved seta.

Moist banks and fields. Spring.

140. **P. rectum,** *Sm.* St. short. L. closely crowded, erecto-patent, elliptic-lanceolate, pointed with an excurrent nerve, often reddish, margins recurved, papillose at back. Caps. exserted, roundish ovoid, on a longish straight seta.

Fields and banks near the coast. Frequent.

Winter and spring.

Fam. 2. Pottieæ.

34. POTTIA, *Ehr.*

A. *Peristome rudimentary or absent.*

1. L. with lamellæ on upper surface.

141. **P. pusilla,** *Hedw.* (*P. cavifolia,* Ehr.). St. very short and simple or branched. L. erecto-patent, concave, obovate or elliptical. Caps. oval, on a short seta, lid obliquely rostrate.

Banks and mud walls.   March.

Var. β. EPILOSA. St. short. L. somewhat acuminate, scarcely piliferous.

Var. γ. INCANA, *N. and H.* L. with long hair-like points.

2. L. without lamellæ.

142. **P. minutula,** *B. and S.* Very minute. L. carinate, spreading, ovate-lanceolate, granulate at back, margins recurved, nerve red, excurrent. Caps. small, ovate-truncate, lid flattish, conical, not beaked.

Fallow fields.   Winter and spring.

Var. β. RUFESCENS. L. narrower, reddish.

Var. γ. CONICA. L. ovate-lanceolate, with a short mucro. Caps. narrower at mouth.

143. **P. truncata,** *L.* St. $\frac{1}{8}$ inch. L. spreading, obovate-acuminate or oblong-lanceolate, broadly concave at base, carinate above, margin plane, with a slightly excurrent nerve. Caps. obovate, truncate, with a wide mouth, lid convex, obliquely rostrate.

Fallow soil.   February, March.

144. **P. intermedia,** *Turn.* Larger than last. L. ovate-lanceolate, margin integrate, revolute below. Caps. oblong, subcylindric, contracted below mouth.

Perist. rudimentary, lid convex, with an oblique beak.

Ditches.

South of England.

145. **P. Wilsoni**, *B. and S.* St. ¼ inch, in tufts. L. larger towards summit, lower ovate-oblong, upper oblong spathulate, obtuse, margin slightly recurved, basal cells lax, hyaline, upper chlorophyllose, verruculose, nerve excurrent into a longish mucro. Caps. elliptic-oblong, contracted at mouth, lid shortly and obliquely rostrate. Calyptra rough at apex.

Sandy banks.   February.

146. **P. crinita**, *Wils.* St. ¼ inch, tufted. L. obovate-oblong, spathulate, obtuse, nerve excurrent into a very long, rigid, hair-like point. Caps. elliptic-oblong, scarcely contracted. Calyptra smooth.

Rocky and moist places.   February, March.

Scotland, Cornwall, Guernsey, Ireland.

147. **P. viridifolia**, *Mitt.*, Journ. of Bot., ix. 4 (*P. pallida*, Braith., Journ. of Bot., viii. 255, non Lindberg). L. obovate-spathulate, obtuse or slightly acute, nerve not very stout, excurrent into a short point, margin recurved at middle, areolæ, upper hexagonal or nearly square, obscure, with minute protuberances, lower oblong, hyaline, smooth. Caps. oblong on a short seta, lid rostrate. Antheridia in axils of comal leaves.

Plymouth (Holmes).

148. **P. littoralis**, *Mitt.*, l.c. L. oblong-spathulate, obtuse or acute, lower pale, upper green, nerve excurrent, longer in lower leaves, areolæ in upper part of leaf small, obscure, smooth, lower oblong, pellucid. Caps. oblong-oval, mouth less than greatest diameter, lid rostrate, slightly twisted. Male fl. bud-like.

Aldington, near Brighton ; Hastings.

149. **P. asperula,** *Mitt.*, l. c. L. obovate-spathulate, acute, but not acuminate, nerve excurrent into a short point, areolæ, upper rounded, rather obscure, each with several elevated points, lower oblong, smooth, pellucid. Caps. oval, lid rostrate, slightly twisted. Antheridia naked, in axils of comal leaves.

Henfield, Sussex ; Penzance (Curnow) ; Jersey (Piquet).

150. **P. Heimii,** *B. and S.* St. ½—¼ inch, cæspitose, branched. L. spreading, oblong-lanceolate, margin not recurved. Caps. obovate or oblong truncate, lid obliquely rostrate, adherent to columella.

Moist banks near the sea.   April, May.

B. Peristome distinct (Anacalypta).

151. **P. Starkeana,** *N. and H.* Minute, gregarious. L. spreading, ovate-lanceolate, entire, papillose above, margin recurved, nerve excurrent. Caps. small, oval, brown, lid convexo-conical. Perist. teeth obtuse, perforate.

Banks and fields.   January, February.

Var. β. BRACHYODUS. Caps. narrower. Perist. teeth very short, truncate.

152. **P. cæspitosa,** *Bruch.* Minute, cæspitose. L. oblong-lanceolate or ovate, concave, plane, nerve excurrent. Caps. ovate, yellowish-brown, lid with a long beak. Perist. teeth perforate.

Woolsonbury Hill, Sussex (chalk).   March.

153. **P. lanceolata,** *Röhl.* St. ¼—½ inch, cæspitose. L. spreading, ovate-lanceolate, acute, margin recurved, entire, nerve excurrent into a longish mucro. Caps. elliptic-ovate, lid conical, obliquely rostrate. Perist.

H

tceth very variable, rather long, with a medial line, strongly papillose.

Moist limestone banks, walls, &c.

35. DIDYMODON, *Br. and S.*

1. Monoicous.

154. D. rubellus, *B. and S.* St. ¼—1 inch, cæspitose. Lower leaves reddish, upper dull green, all oblong-lanceolate, spreading, papillose on both sides, margin recurved, keeled, nerved nearly to apex. Caps. pale brown, cylindrical, lid with a short, oblique beak. Antheridia naked, in axils of per. l.

Shady walls, rocks, banks, &c.   October.

2. Dioicous.

a. L. lanceolate, rigid.

155. D. luridus, *Hornsch.* St. ¼—1 inch, cæspitose. L. lower ovate-lanceolate, upper oblong-lanceolate, with entire recurved margins, keeled, acute, nerved (reddish) almost or quite to apex, areolæ small, roundish. Caps. symmetrical, oblong, on a shortish seta twisted to the right, lid conical, pointed. Perist. teeth small, irregular.

Limestone walls, &c.   Rare.   December.

b. L. narrow, not rigid.

156. D. flexifolius, *Hook. and Tayl.* Barren stems long, trailing, fertile, ½ inch. L. spreading, flexuose, more so when dry, oblong or ligulate, margin reflexed below, and serrate at apex, nerve not reaching apex, areolæ round. Caps. small, cylindrical, somewhat curved, lid with a short beak. Perist. teeth short.

High moorlands.   March, April.

Buxton, Alderley Edge, Ben Ledi, &c.

Var. GEMMESCENS, *Mitt. MS.* Nerve excurrent into

an apiculus, which bears a cluster of egg-shaped or oblong gemmæ.

Old thatch.

Amberley, Sussex (Mitten), &c.

157. **D. cylindricus,** *B. and S.* ¼—1 inch. L. spreading, flexuose, linear-lanceolate, margin undulate and minutely crenulate, nerve narrow, forming apiculus, areolæ small, opaque, gradually enlarged towards the base, there diaphanous. Caps. erect, narrow, cylindrical, lid long, conico-rostrate. Perist. teeth linear-lanceolate, fugacious.

Damp shady rocks.   October.

158. **D. sinuosus** (*Schp.*), *Mitt.* (*Dicranella*, Wils. MS. ; *Trichostomum*, Lindb.). Densely cæspitose, fuscous below. L. long, linear-lanceolate or subulate, patent, from a very short, pellucid base, margin slightly recurved below, above denticulate, nerve continued into a thick, obscure, blunt point, often broken off, basilar cells all oblong and rectangular. Fr. not known.

Shady places at roots of trees.

Sussex, Cornwall, Bangor.

159. **D. recurvifolius,** *Tayl.* "Stems elongate, loosely cæspitose. L. squarrose, crisped and undulate when dry, elliptic, oblong or ligulate, pale margined, serrulate, nerve subexcurrent, areolæ small, dense, opaque, elongate and pellucid at base."

Ireland, 1842 ; fruit not known (Wilson).

Fam. 3. **Trichostomæ.**

36. **TRICHOSTOMUM,** *Br. and S.*

160. **T. tophaceum,** *Brid.* St. ¼—1 inch, densely cæspitose. Br. fasciculate. L. lanceolate (the upper ones obtuse), concave, keeled, margins recurved, nerve

not reaching to apex.   Caps. subcylindrical, erect, regular, lid with an oblique beak.   Perist. teeth variable, sometimes only 16, somewhat fugacious.

Moist places and rocks.   November.

161. **T. brachydontium,** *Brid.* (*T. mutabile,* Bruch.). L. broader, lanceolate or ligulate, crisped, not cucullate, margin slightly undulate, nerve excurrent into a prominent mucro.   Caps. ovate-oblong, erect, regular, lid obliquely rostrate.   Perist. t. very short and irregular.

Damp rocks and fissures.   April—June.

162. **T. crispulum,** *Bruch.*   St. ¼—1 inch.   L. lower lanceolate, distant, upper crowded, longer, linear-lanceolate, concave, cucullate at apex, crisped when dry, nerve prolonged into a short mucro.   Caps. oval, erect, regular, lid with an oblique beak.   Perist. teeth in unequal pairs.

Limestone rocks near the sea.   June, July.

Ormes Head, Anglesea, Bristol, &c.

Var. ε. ELATUM.   Larger and more robust.   L. longer, apex cucullate, muticous.

Ireland, Somerset (H. Boswell).

163. **T. flavo-virens,** *Bruch., Muller.*   St. short, with innovations from summit (interruptedly comose).   L. glaucous or yellowish-green, oblong-ligulate, obtuse, mucronate, margins entire, undulate, incurved, thick nerve prolonged into a short mucro.   Fruit-stalk red at base, yellowish above, slightly flexuose.   Caps. oblong, cylindrical, pale yellowish-brown, with a red mouth.   Perist. t. elongate, regular, in pairs, lid acuminate, half as long as caps., with an oblique beak.

Shoreham, Sussex; Plymouth (Holmes); Malahide (Dr. Moore); &c.   Spring.

164. **T. nitidum,** *Lindb.,* 1864 (*T. diffractum,* Mitt.,

1868). Densely pulvinate, olive-green above, brown below. St. rigid, branched. L. crowded, erecto-patent, arcuate when dry, more or less elongate, linear or sublingulate, channelled, margin plane, slightly undulate, nerve terete, prominent on back, excurrent, areolation minute, loose and cuneiform at base, papillose above. Fruit not known.

Clifton, Torquay, Plymouth, &c.

165. **T. littorale,** *Mitt.* St. elongate, tufted, more or less interruptedly comose. L. patent from an erect base, oblong-ligulate, obtuse, concave, recurved towards apex, with nerve excurrent into a short mucro, basal cells hyaline, oblong, and rectangular.

Ireland; Whitsand Bay, Cornwall; Hastings; Devonshire; &c.

37. BARBULA, *Hedw.*

Section 1. *ALOIDELLA.*

L. rigid, nerve broad, bearing articulate filaments on its upper surface.

166. **B. brevirostris,** *B. and S.* (*Muell.* ?). Plants minute. L. lower rotundate-ovate, upper broadly oblong, obtuse, nerve thin, annulus broadish. Perist. t. once convolute. Monoicous. (Sch. Syn., ed. 2, 189.)

Clay soil. July, August.

Near Edinburgh (Stewart), near Buxton (E. George, 1873).

167. **B. stellata,** *Schreb.,* 1771 (*T. rigida,* Schultz). St. minute, loosely cæspitose. L. spreading from an upright base, oblong, obtuse, margin inflexed membranaceous. Caps. erect, ovate-elliptical, lid with a long oblique beak. Calyptra half as large as capsule. Perist. teeth long, and much twisted. Dioicous.

Limestone walls.   October—March.

168. **B. ambigua,** *B. and S.*   Larger in all its parts
than last.   L. ligulate, lanceolate, apex cucullate,
margin incurved.   Caps. erect, cylindrical, lid rostrate.
Calyptra very short.   Perist. teeth filiform, little
twisted, arcuato-incurved when dry.   Dioicous.

Walls and banks (marly).   November, December.

169. **B. aloides,** *Koch.*   St. as above.   L. spreading,
narrowly lanceolate, acute, with a strong nerve.   Caps.
elongate-cylindrical, inclined, lid conical, acutely rostrate.   Perist. teeth scarcely twisted, when dry widely
spreading.   Dioicous.

Clay banks.   Winter and spring.

170. **B. lamellata,** *Lindb.* (*Pottia cavifolia* var. *gracilis*, Bry. Brit.   *B. cavifolia*, Schp. Syn.).   St. very
short, cæspitose.   L. rather lax, erecto-patent, concave,
lower smaller, roundish oval, piliferous, upper larger,
oval, spathulate, nerve excurrent into mucro.   Caps.
oblong, subcylindrical, striate when dry, on a long red
seta, lid with a long, rather oblique beak.   Perist. that
of a true *Tortula*, but so fragile as to have escaped
notice, and always falling off with the operculum.
Monoicous.

Banks and walls.   February.

Oxford (Boswell), Pontefract, Edinburgh (Nowell),
Aldrington (Davies), &c.

Section 2.   *CUNEIFOLIÆ.*

L. broadly or spathulato-lanceolate.

171. **B. atrovirens,** *Smith, Lindb.*   [*Didymodon nervosus*, Hook. and T.   *Desmatodon nervosus*, Bry. Brit.]
St. ¼ inch, densely cæspitose, branched.   L. spreading,
oval, spathulate or oblong, concave, margins revolute,

nerve thick, prolonged into a short mucro, areolæ small, roundish, larger and diaphanous at base, slightly papillose. Caps. oval-oblong, lid large, conical, with an oblique beak. Monoicous.

Dry banks, &c., near the sea. Spring.

172. **B. cuneifolia,** *Dicks.* Gregarious. St. simple. L. upper crowded, broadly obovate, acuminate, nerve sometimes excurrent, soft, pellucid, areolæ loose, lower broadly ovate, aristate. Caps. elliptic, erect, lid obtuse, short. Basilar membrane of peristome broadish. Monoicous.

Banks, seacoast. Rare. March, April.

173. **B. Vahliana,** *Schultz.* Small, gregarious or cæspitose. L. lower oblong, upper oblong wedge-shaped, nerve excurrent, subulate, margin erect, sometimes reflexed, crenulate. Caps. narrow, elongate, cylindrical, brown, sometimes slightly incurved, lid shortly subulate, annulus broad. Basilar membrane of perist. tesselate. Monoicous. Differs from *muralis* in its broader, softer leaves, narrower capsule, and longer basilar tube.

Damp clayey ground, on roadsides, &c. Spring.

First discovered at Maresfield in Sussex, 1863, by G. Davies; Woking, Surrey (Sheppard and Westell); Dublin (Moore); &c.

174. **B. marginata,** *B. and S.* St. simple, gregarious or cæspitose. L. oblong-lanceolate or linear, concave, margin thickened, nerve sometimes excurrent into a mucro. Caps. cylindric-ovate or oblong, lid very large, shortly rostrate. Basilar membrane narrow.

Sandstone walls. Rare. May, June.

Sussex, I. of Wight, Cheshire, Yorkshire, &c.

175. **B. canescens,** *Br.* Simple, gregarious or cæspi-

tose, hoary. L. lower obovate, upper oval-oblong or oblong-lanceolate, minutely papillose above, all concave, with a recurved margin, and nerve excurrent into a long, hair-like point. Caps. small, oblong, erect, with a long, oblique, conical lid. Basilar membrane a broadish, tesselated tube. Monoicous.

Fairlight Glen, Hastings (Mr. Jenner). Spring.

176. **B. muralis,** *Timm.* Short, cæspitose. L. oblong, obtuse, margin recurved, apex unequal or subcordate, nerve excurrent into a long, hair-like point. Caps. oblong or subcylindric, erect, with a long, rostellate lid. Basilar membrane narrow. Monoicous.

Var. β. INCANA. Caps. small. L. oval-lanceolate, with long hair-points.

Var. γ. ÆSTIVA. L. long, linear-lanceolate, nerve scarcely excurrent.

Var. δ. RUPESTRIS. Larger and much branched. L. larger, oblong, piliferous. Caps. longer, curved.

Walls and stones (δ limestone). April, May.

Section 3. *TORTULA.*

St. more or less elongate. Br. dichotomous. L. larger above, more or less elongate lanceolate; per. l. sheathing. Perist. teeth long, much twisted.

a. Unguiculatæ.

L. oblong, linear lanceolate, nerve scarcely excurrent into a mucro. Perist. teeth with basilar membrane short.

177. **B. unguiculata,** *Hedw.* St. ⅛—1 inch, cæspitose, dichotomous. L. oblong-lanceolate, obtuse, margin recurved below, densely papillose above, nerve excurrent into a short mucro; per. l. hyaline, with a long excurrent nerve. Caps. oblong, cylindrical, erect,

on a long, reddish seta, lid with a subulate beak. Dioicous. .

Clay banks and hedges. Spring.

Var. β. CUSPIDATA. St. shorter. L. narrower, with a longer mucro.

Var. γ. APICULATA. L. spreading, recurved, mucro long.

178. **B. fallax,** *Hedw.* St. ½—1 inch, cæspitose. L. lanceolate from a broadish base, keeled, margin recurved, slightly papillose, somewhat squarrose, gradually tapering and nerved to apex; per. l. sheathing halfway, thence narrowly lanceolate, patent. Caps. variable both in size and shape, usually subcylindrical, with an obtuse, rostrate lid, often as long as itself. Dioicous.

Clay and limestone banks and walls.

October—December.

Var. β. BREVICAULIS, *Schw.* St. short, simple. L. closely set, patent, margins subundulate. Caps. shorter, as also the perist. and lid.

Parapet of bridge, Earlswood, Warwick (J. E. Bagnall, Dec., 1883).

Var. γ. BREVIFOLIA. St. short, simple. L. crowded, patent, margin subinvolute. Caps. shorter.

179. **B. recurvifolia,** *Schp.* [*reflexa,* Brid.; *T. fallax,* δ. Bry. Brit.] Tufts always of a ferruginous tint. St. loosely cæspitose. L. tristichous, recurved, and falcate, slightly twisted, from an oblong wide base, lanceolate, keeled, strongly papillose on both sides, margin reflexed below, nerve brownish, vanishing below apex. Caps. erect, cylindrical, regular, lid subulate, beaked. Dioicous.

Calcareous rocks and walls. Rare in fr. Autumn.

Scotland; Yorkshire; Derbyshire; Rydal Water
(Baker); &c.

180. **B. rigidula,** *Hedw.* (*Trichost. rigidulum,* var. β.
densum, Bry. Brit. 114.). L. patent, lanceolate, carinate, margin recurved, rigid, bristly, not appressed
and imbricate, when dry slightly curved and loosely
contorted, nerve stout, continued into a thick, obscure
point, not really excurrent. Caps. cylindric, erect, on
a reddish seta, lid obliquely rostrate.

Scotland, York, Sussex, Cornwall, &c.

Spring and summer.

181. **B. Woodii,** *Schp.* Yellow-green above, ferruginous below, radiculose. L. spreading or even slightly
recurved, twisted when dry, narrowly linear-lanceolate,
subulate, concave at base, one or other margin recurved,
nerved (yellowish) to or sometimes beyond apex, cells
minutely quadrate above, oblong-rectangular at base.
Fr. unknown.

Cromagloun, Killarney, July, 1865, Schp.

182. **B. spadicea,** *Mitt.* (*T. rigidulum,* Smith, Bry.
Brit., p. 114). St. robust, 1—2 inches. L. patent
from the base, lanceolate-subulate, canaliculate, margin
recurved below, incurved and closely imbricate when
dry, nerve percurrent and distinct to apex; per. l.
lower half erect, broadly ovate, upper narrow, recurved.
Caps. erect, cylindrical, on a red seta, lid shortly
subulate, twisted. Teeth narrow, on a short membrane. Dioicous. (Mitten, Journ. of Bot., v. 326.)

Rocks and stones near water.   Autumn, winter.

Scotland, Ireland, Bolton Abbey, &c.

183. **B. cylindrica,** *Tayl.* (*T. insulana,* De Not.; *T.
rincalis* β. *flaccida,* Bry. Brit., p. 124). St. loosely
caespitose. L. lower lanceolate, upper lincar-subulate,

from a lanceolate, appressed base, patent, spreading, more or less arcuate, papillose, acute, margin recurved below, above plane, contorted when dry, nerve excurrent. Caps. cylindric or elliptic, erect or slightly curved, lid conical, attenuate, somewhat obtuse, half as long as capsule. Rare in fruit.

England and Ireland. Frequent. May.

184. **B. vinealis,** *Brid.* Tufts more robust. L. patent or subrecurved, from a broad base, elongate and narrowly lanceolate, nerve broad, obscure towards the acute apex, the latter usually tipped with a pointed hyaline cell, appressed when dry, not crisped or contorted. Caps. subcylindrical. Dioicous.

Walls. April, May.

b. Convolutæ.

L. bright green, incurved, twisted when dry, cells lax at base; per. l. much sheathing and convolute. Caps. small, ovato-elliptical, slightly curved. Perist. t. long, contorted.

185. **B. Hornschuchiana,** *Schultz.* Loosely cæspitose. L. crowded, spreading, ovate-lanceolate, carinato-concave, obsoletely papillose, gradually tapering to an acute point, formed by slightly excurrent nerve, margin slightly revolute. Caps. oblong, slightly curved, annulus narrow, seta red below, yellow above. Dioicous.

Rocks, walls, and banks. Not common. April, May.

[186. **B. paludosa,** *Schwg.* St. radiculose. L. erecto-patent, linear-lanceolate, acute, acutely carinate in the upper half, margin plane, but generally slightly serrulate at apex ; per. l. sheathing, distinctly serrate. Caps. ovate-oblong, often curved, on a flexuose seta, reddish below, lid as long as the caps. Perist. t. very long, much twisted.]

Bogs and damp places.   Autumn.

Recorded to have been gathered by Mr. Mitten in
Wales, August, 1883, but some of the specimens dis-
tributed are certainly *Trichost. brachydontium.*

187. **B. revoluta,** *Schwaeg.* Densely cæspitose. L.
crowded, erecto-patent, oblong lanceolate, obtuse, nerve
excurrent from the blunt apex, margin strongly revolute.
Caps. reddish brown, elliptic, symmetric, with a scarcely
oblique lid, seta red, straw-coloured above. Dioicous.

Walls, mostly limestone.   May.

188. **B. convoluta,** *Hedw.* Cæspitose. L. spreading,
lower ovate-lanceolate, upper narrowly ligulate, acute,
lanceolate, plane or somewhat undulate in margin,
which is reflexed towards base, nerve not excurrent ;
per. l. strongly convolute, nerveless, sheathing. Caps.
inclined, oblong-ovate, on a yellowish seta, lid with a
long, oblique beak. Dioicous.

Walls, &c.   May, June.

189. **B. commutata,** *Jurat.* Tufts green above, rufescent below, taller and more robust than the last.
L. spreading, subrecurved, densely and minutely
papillose, lower ovate-lanceolate, middle lanceolate,
upper lanceolate from an ovate-concave, sheathing
base, acute, carinate, margin reflexed, slightly undulate, and minutely crenulate with papillæ, nerve thick,
vanishing at apex. Caps. narrowly oblong, incurved
on a twisted yellow seta, lid half as long as capsule or
more conical, subulate.

Scotland, Ireland. Rare.

c. Tortuosæ.

Tufts large, robust. L. linear-lanceolate, flexuose,
margins undulate, scarcely recurved, crisped, and
twisted when dry. Perist. t. with scarcely a basal
membrane.

a. Dioicous.

190. **B. inclinata,** *Schwg.* Tufts dense, robust, densely leaved. L. elongate-linear, broadly undulate, whitish, nerve excurrent into a mucro; per. l. narrower, longer, erect, laxly areolate. Caps. oval-oblong, more or less incurved, gibbous at base, on a red, spirally twisted seta, lid narrowly conical.

Sandy soil near river beds. Spring.
Oxfordshire (H. Boswell).

191. **B. tortuosa,** *W. and M.* St. ½—3 inches, tufted. L. very long, linear-lanceolate, crowded, flexuose, margin plane and undulated, with an excurrent nerve; per. l. narrow and tapering, cirrhate. Caps. straight or incurved, erect or inclined, ovate-oblong, on a longish seta. Dioicous.

Limestone rocks. July.
Derbyshire.

192. **B. Hibernica,** *Mitt.* (*B. cirrifolia*, Schp.). St. 2 inches, branched. L. at apices of branches subcomose and stellate, patent or patent-divergent, straight, rarely incurved or recurved, concave, cirrhate when dry, from an oval oblong, semi-amplexicaul base, longly lanceolate, subulate, margin plane, minutely crenulate, acute, nerve yellow, continued to apex or beyond.

Mountains near Dunkerran, common, but always sterile (Dr. Taylor). Cromagloun, Killarney.

193. **B. fragilis,** *Wils.* (*Trichostomum*, Müll. Syn.). St. erect, simple or dichotomously branched, radiculose, tomentose at base. L. crowded, lanceolate-subulate, nerve excurrent, margins plane, crenulate, areolæ minute, large and hyaline at base, papillose. Caps. erect, ovate-oblong, subcylindric, regular or slightly incurved, lid conical, with a long, oblique beak. Fr. rare.

Clefts of rocks and on the ground.   Summer.
Ben Lawers.

194. **B. squarrosa**, *De Not.* St. 1 inch, cæspitose.
L. squarrose, lanceolate, recurved, with a broad,
sheathing base, margin undulate, with large, diaphanous
cells, somewhat serrulate at apex, nerve scarcely excurrent. Caps. sub cylindrical, narrow, slightly curved,
lid conical, half as long as capsule, seta 1 inch long.
Dioicous.

Chalk. Ireland and S. of England. Fruit not
known in this country.

195. **B. Brebissoni**, *Brid.* [*Cinclidotus riparius β.
terrestris*, Bry. Brit.] St. 1—2 inches, radiculose.
Branches fastigiate. L. erecto-patent, long, lingulate,
concave, minutely papillose on both sides, margin
slightly recurved, nerve thick, excurrent into a mucro,
areolæ upper minute, rectangular and hyaline at base.
Caps. erect, cylindrical, incurved, lid with an oblique
beak. Dioicous.

Stones in streams.   April.
Anglesea, Bristol, Surrey, Sussex, in fr. (Davies), &c.

d. Syntrichia.

Plants robust. L. oblong-lingulate or ovate-spathulate, chlorophyllose and papillose above, smooth and
hyaline below. Caps. erect, oblong or cylindric, subarcuate. Perist. teeth long, much twisted.

1. *Subulatæ.*

196. **B. subulata**, *Brid.* Cæspitose, simple or
branched. L. obovate or spathulate oblong, narrowed
at base, margin plane, sometimes with a row of larger
cells, nerve excurrent into a short mucro, apex sometimes slightly toothed. Caps. very long, cylindrical,

curved, with a short lid. Half peristome tubular. Monoicous.

Sandy hedge-banks, walls, &c.   May, June.

197. **B. lævipila,** *Brid.* Cæspitose. L. spreading, often recurved above, obovate oblong or almost panduriform, margin slightly recurved below, nerve reddish, excurrent into a longish white hair-point from the obtuse apex. Caps. cylindrical, slightly curved, lid conical. One-third peristome tubular. Monoicous.

Trunks of trees and rocks.   May, June.

198. **B. latifolia,** *B. and S.* L. obovate-spathulate or almost panduriform, soft and flaccid, with a scarcely excurrent nerve, notched at the obtuse apex. Caps. cylindrical, slightly curved, with a long, rostrate lid. Quite one-third of peristome tubular. Annulus small. Dioicous.

Roots of trees, stones, &c. Fruit rare.   Summer.

199. **B. ruralis,** *Hedw.* Cæspitose. Branches dichotomous. L. squarrose, recurved, ovate-oblong, keeled, nerve excurrent into a long, scabrous hair-point, from the obtuse apex, margin slightly recurved. Caps. subcylindrical, slightly curved. Quite one-half perist. tubular, lid long, conical. Dioicous.

Walls and roofs.   March, April.

200. **B. intermedia,** *Brid.* (*T. ruralis β. minor,* Wils. Bry. Brit.). Smaller and densely cæspitose or subpulvinate. L. erecto-patent, oblong, spathulate, apex obtuse, nerve excurrent into a long, scabrous hair-point. Caps. shorter than in last. Dioicous.

Limestone walls.   Spring.
England, Scotland, N. Wales, &c.

201. **B. papillosa,** *Wils.* Cæspitose. L. spreading, obovate, concave, margin plane (involute when dry),

nerve thick, bearing short articulate threads above,
papillose on the back, and excurrent into a smooth
hair-point from suddenly tapering apex, a few hyaline
cells at base. Fr. not known.

Trunks of trees, &c.

202. **B. princeps,** *De Not.* (*T. Mulleri,* B. and S.).
St. 1—2 inches, cæspitose, with brownish radicles.
L. erecto-patent, oblong, broad, concave, fawn-coloured,
margin reflexed, nerve excurrent into a short, scabrous
hair-point, from a rounded, obtuse apex. Caps. cylindrical, straight or curved, on a purplish seta. One-half
peristome tubular. Synoicous.

Rocks.   Summer.

Scotland, Ireland.

Tribe xi. CALYMPERACEÆ.

38. **ENCALYPTA,** *Schreb.*

a. Peristome absent.

203. **E. commutata,** *N. and H.* St. about 1 inch,
branched, radiculose.   L. squarrose, lower ovate-
lanceolate, upper elongate-acuminate, concave, acute,
transversely plicate, nerve excurrent. Caps. smooth,
cylindrical, with a long beaked lid. Calyptra jagged,
but not fringed at base, longer than capsule, brown at
apex. Monoicous.

Alpine summits.   July, August.

Scotland.

b. Peristome single.

204. **E. vulgaris,** *Hedw.* St. about ½ inch, branched,
radiculose. L. spreading, elliptic-lanceolate, oblong,
acute or obtuse, subundulate, apiculate, nerve sometimes
excurrent, margin plane. Caps. smooth, cylindrical.
Base of calyptra entire. Perist. very fugacious of 16
teeth. Monoicous.

ENCALYPTA. 113

Limestone walls, rocks, &c.   March, April.

Var. β. Perist. none. L. apiculate (common).

Var. γ. Perist. none. L. obtuse and concave at apex.

Var. δ. Perist. none. L. obtuse. Caps. oblique.

Var. ε. Perist. none. L. piliferous.

205. **E. rhabdocarpa,** *Schw.* St. ½—1 inch, radiculose. L. spreading, erect and crisped when dry, oblong-lanceolate, nerve generally more or less excurrent into a mucro, margins crenulate with papillæ above. Caps. narrowly ovate or ovate cylindric, striate, ribbed when dry. Perist. persistent. Calyptra slightly toothed at base, roughish at apex, yellow. Monoicous.

Mountains in Scotland and Ireland. July, August.

206. **E. ciliata,** *Hedw.* St. about ½ inch, radiculose. L. oblong-ovate, margin recurved below, and toothed near apex, strongly verruculose, gradually tapering to a point formed by the excurrent nerve, undulate. Caps. cylindrical, elongate, smooth. Perist. persistent. Calyptra fringed at base. Monoicous.

Subalpine rocks.   June, July.

c. Peristome double.

207. **E. streptocarpa,** *Hedw.* St. 1—2 inches, radi-culose. L. suberect, ligulate, obtuse and cucullate at apex, nerve not excurrent; per. l. lanceolate-subulate from an ovate base. Caps. oblong, narrowed above, spirally striate, and twisted when dry. Perist. outer teeth filiform, inner cilia. Calyptra toothed or fringed at base, and roughened at apex. Dioicous.

Limestone and mortared walls. Rare in fruit.

August.

I

Tribe xii. GRIMMIACEÆ.

Fam. 1. **Grimmieæ.**

39. **GRIMMIA,** *Ehr.*

Section 1. *SCHISTIDIUM.*

Caps. immersed. Calyptra mitriform, lobed.

208. **G. conferta,** *Funk.* Cæspitose, intense green above, blackish below. L. ovate-lanceolate, tapering in the upper ones to a short hair-point, margins slightly recurved and thickened, nerve strong, deeply channelled on its upper side. Caps. small, ovate, with a rostellate lid, almost pellucid. Perist. teeth much perforated, pale or orange-red.

Rocks.   February, March.
Scotland.

Var. β. URCEOLARIS. Caps. urceolate. L. with white points.

Var. γ. OBTUSIFOLIA. L. all obtuse, shorter and broader.

Var. δ. INCANA (*G. pruinosa,* Wils. MS.). More robust, per. l. broader, with long hair-points. Caps. more elongate. Perist. teeth stronger, nearly entire, red. [Dr. Braithwaite, Journ. Bot. (N.S.), vol. i. 195.]

Trap rocks.

King's Park (Greville); Arthur's Seat, Edinburgh (Bell); Fife (Howie).

209. **G. apocarpa,** *L.* Loosely cæspitose. L. spreading, lanceolate acuminate, from an ovate erect base, with white points, upper ones with a short, rough hair-point, margins much recurved, nerve ceasing below apex; per. l. larger, with a thinner nerve. Caps. ovate, thick-walled, with an oblique beaked lid. Perist. teeth dark red. Calyptra divided at base.

Rocks and walls, sometimes on trees.

November—March.

Var. β. GRACILIS. Per. l. secund, others subsecund or spreading, stem decumbent, elongated.

Var. γ. RIVULARIS. St. fasciculate. L. ovate-lanceolate, dark green, obtuse. Caps. turbinate.

By streams.

210. **G. maritima,** *Turner.* Cæspitose, dull green or brownish. L. rigid, not hair-pointed, straight, lanceolate acuminate, keeled, nerve strong, reddish-brown, excurrent, margin plane. Caps. obovate, with a large, rostellate lid. Perist. teeth large and perforate.

Rocks near the sea. November—March.

Section 2. *GASTERO-GRIMMIA.*

Plants very short, pulvinate. Caps. slightly emerging, ventricose on one side, on a short, curved seta. Calyptra five-lobed or cucullate.

a. Peristome absent.

211. **G. anodon,** *B. and S.* In small, hoary cushions. L. lower minute, loosely imbricate, ovate lanceolate, muticous, upper larger, broadly oblong-lanceolate, concave, nerve excurrent into a long serrated hair, basal cells elongate, pellucid, above quadrate, opaque. Caps. immersed, oval, strongly ventricose, lid plano-convex. Monoicous.

Walls and dry limestone rocks. Spring.
Arthur's Seat (Bell).

b. Peristome perfect.

212. **G. crinita,** *Brid.* In loose, flat, silky tufts. L. lowest imbricate, lanceolate, muticous, upper obovate-oblong, strongly concave, the broad, diaphanous

apex continued into a long hair, nerve not reaching apex, margin plane, basal cells elongate, diaphanous, upper large, rounded, thickened. Caps. ovate, lightly striate, subcernuous, furrowed when dry, lid convex, with an obtuse point. Calyptra dimidiate, two-lobed. Monoicous. [Dr. Braithwaite, Journ. Bot. (N.S.), vol. i. 195.]

Mortar of old walls and limestone rocks.

Autumn, spring.

Near Hatton, Warwick, 1872 (J. Bagnall).

Section 3. *EU-GRIMMIA.*

Caps. exserted, regular, on an arcuate seta. Calyptra mitriform, lobed, rarely cucullate.

a. Calyptra cucullate.

213. **G. orbicularis**, *B. and S.* Densely pulvinate. L. oblong lanceolate, rounded obtuse at apex, with nerve excurrent into a long hair-point, basal cells large. Caps. almost spherical, drooping, on a curved yellowish seta, slightly striate, lid small, convex, annulus narrow. Perist. teeth trifid, more distantly barred than the next. Monoicous.

Limestone rocks.   February, March.

b. Calyptra lobed, upright or oblique.

1. Monoicous.

214. **G. pulvinata**, *Dill.* Densely pulvinate. St. ½—1 inch. L. elliptic, lanceolate, margin recurved, apex rather obtuse, terminated by the nerve excurrent into a long hair-point. Caps. drooping, reddish-brown, ovoid, eight-furrowed, lid convex, with a straight beak. Calyptra lobed at base. Perist. teeth dark red, bitrifid, annulus large.

Rocks and walls. April, May.

Var. β. OBTUSA. Lid short, obtuse. Caps. shorter.

215. **G. Schultzii,** *Brid.* L. crowded, subsecund, olive-green, black when dry, from an oblong base, elongate, lanceolate, tapering into a long, rough, diaphanous point, margins recurved. Caps. slightly obovate, eight-furrowed, on a very short, curved seta, annulus large. Perist. teeth long, tapering, deeply bifid. Monoicous.

Subalpine rocks. April, May.

England, Scotland, Wales.

2. Dioicous (or fruit not known).

216. **G. robusta,** *Fergusson MS.* In large, loose tufts, black below, dark green and hoary above. Branches fastigiate. L. erecto-patent, appressed when dry, keeled at back with the strong nerve, margin recurved below, lower short, muticous, lanceolate from a contracted ovate base, upper longer, gradually tapering into a long, smooth hair point, cells quadrate, thickened, at centre of base longer, with a single row at margin of basal wing hyaline. [Dr. Braithwaite, l.c., p. 196.]

Alpine rocks.

Clova (Fergusson); Fairhead, Ireland (Dr. Moore); Cardross and Bowling (Dr. Stirton); Ross-shire (Hunt).

217. **G. contorta,** *Wahl.* In small, deep green, soft tufts, black below and radiculose. L. patent, incurved, curled when dry, lineal subulate from a lanceolate base, with short, diaphanous hair-points, keeled, margin recurved below, basal cells diaphanous, elongate, hexagono-rectangular, above sinuous and quadrate; per. l. erect, sheathing. Caps. small, oval, smooth, yellowish, cernuous on a subarcuate seta, erect when

dry, lid convex, conical, obtuse, orange-red. [Dr.
Braithwaite, l.c., p. 197.]

Quartz rocks.

Cheviots (Hardy) ; Cloch-na-ben (Sim) ; Glen Callater, &c. (Fergusson).

218. **G. torquata,** *Grev.* (*G. torta,* N. and H., Bry.
Brit.). Loosely tufted, elongate. St. 1—2 inches,
dichotomous, radiculose only at base. L. erecto-patent,
lanceolate-acuminate, spirally twisted when dry, channelled, upper hair-pointed, lower muticous, channelled,
margin plane. Fruct. not known.

Alpine rocks.

England, Scotland, Ireland.

219. **G. funalis,** *Schwgn.* (*G. spiralis,* H. and T.,
Bry. Brit.). Densely pulvinate. St. ½—1 inch, slender.
L. oblong or ovate-lanceolate, erecto-patent, upper ones
tapering into a long hair-point, nerve not excurrent.
Caps. ovoid, smooth, eight-furrowed when dry, lid
short, apiculate, orange-coloured, annulus large, compound. Calyptra five-lobed at base. Perist. teeth
closely bifid, lacunose at apex.

Dry alpine rocks.   August—November.

England, Scotland, Ireland.

220. **G. Muhlenbeckii,** *Schpr.* Loosely pulvinate and
cæspitose. St. tall, erect or procumbent, dichotomous,
and rooting at base. L. densely crowded, patulous,
erect when dry, elongate-lanceolate, keeled with the
strong nerve, margin plane, lower with a short, upper
with a long, rough hair-point, with recurved teeth,
basal cells elongate, upper rounded, quadrate. Caps.
small, oval, glossy, rugulose when dry, yellowish-brown,
lid convex, with a short beak, red. [Dr. Braithwaite.
l.c., p. 197. Schp. Syn., p. 255.]   July.

221. **G. subsquarrosa,** *Wils. MS.* Dr. F. B. White, Bot. Soc. Edin. Trans., ix. 142. In lax, dark green tufts, fuscous at base. St. $\frac{1}{3}$—$\frac{3}{4}$ inches, with dichotomous, short, curved branches. L. patent, squarrose, erect and appressed when dry, lowest from an ovate base, gradually lanceolate, muticous, upper longer, and extended into a long, denticulate hair-point, nerve strong, margin recurved, basal cells quadrate, hyaline, marginal narrow and elongate, above minute, rounded quadrate. Fr. not known. [Dr. Braithwaite, l.c., p. 196.]

Rocks.

Kinnoul Hill, Perth (Dr. B. White); Moncrieff Hill (Dr. Stirton); Arthur's Seat and Braid Hills, &c. ; near Radnor, 1874 (H. Boswell).

222. **G. trichophylla,** *Grev.* Loose, yellowish-green tufts, $\frac{1}{4}$—1 inch. L. linear-lanceolate, from an erect base, flexuose, upper oblong lanceolate, nerve excurrent, tapering into a long, diaphanous point, margin recurved at base, cells shortly sinuous-rectangular, marginal quadrate. Caps. ovate-oblong, furrowed when dry, lid with a long, straight beak, annulus larger. Perist. teeth bifid. Calyptra lobed. Dioicous.

Walls.   April—June.

England, Scotland, Ireland.

223. **G. Hartmanni,** *Schp.* Loosely cæspitose, green above, black below. St. elongate, procumbent, rigid, arcuate, ascending, dichotomous. L. elongate-lanceolate, upper ones secund, prolonged into a short, smooth hair-point, somewhat concave, margin more or less recurved, basal cells sinuoso-rectangular, hyaline, above quadrate, opaque. Fruit not known. [Dr. Braithwaite, l.c., p. 197. Sch. Syn., 258.]

Shaded quartzose rocks.

Wales and Scotland.

224. **G. elatior,** *Schp.* Robust, loosely cæspitose, fuscous green, hoary at top. St. sparingly branched, elongate, from decumbent naked base ascending. L. very long, curved, patent, from oblong carinato-concave base longly lanceolate, ending in a long, smoothish hair-point, margin revolute, basal cells linear-rectangular, wider towards margin, above rounded, opaque. Caps, ovate, ten-ribbed, when dry oblong, deeply furrowed, lid conical, muticous or subaciculate. Dioicous. [Dr. Braithwaite, l.c., 197. Schp. Syn.. 258.]

Granite rocks.   Spring.

Clova, 1868 (Fergusson).

Section 4. *GUEMBELIA.*

Calyptra mitriform or cucullate. Caps. on an upright seta.

a. Calyptra lobed, mitriform.

1. Monoicous.

225. **G. Donniana,** *Sm.* St. ¼—½ inch, tufted. L. erecto-patent, lanceolate-elongate, narrow, tapering into a long, scarcely roughened hair-point, thinly nerved, margin plane; per. l. longer. Caps. erect, oval-oblong, slightly exserted, pale yellowish-brown, lid obtuse, conical, annulus small. Perist. teeth broad, sometimes perforate.

Mountain rocks and walls.   March, April, October.

England, Scotland, Wales.

Var. β. SUDETICA. L. with longer hair-points. Caps. immersed, lid conico-acuminate.

Var. γ. ELONGATA. L. scarcely hair-pointed. Caps. on a longish seta.

226. **G. ovata,** *W. and M.* St. ½ inch or more, branched, fastigiate. L. spreading, erect when dry, from an oblong, concave base, lanceolate, tapering into a roughish hair-point, nerve broad, indistinct, margin recurved below, basal cells scarcely sinuous. Caps. ovoid, erect, exserted, reddish-brown, annulus large, lid obliquely rostellate, with a groove round its base. Perist. teeth narrow, cleft and perforate.

Alpine rocks.   October—March.

Breadalbane and Clova, Snowdon, Charnwood Forest, &c.

2. Dioicous.

227. **G. leucophea,** *Grev.* Dark green, hoary tufts. St. ½ inch. L. erect, spreading, when dry closely imbricate, upper ovate or elliptical concave, with very long hair-points and plane margins, lower ones smaller, muticous. Caps. smooth, elliptical or oblong, erect, exserted, with a short conico-rostellate lid, and broad annulus.

Scotland, Devon, &c.   April.

b. Calyptra cucullate, dioicous.

228. **G. commutata,** *Hueb.* Loosely tufted, blackish-green, hoary at top. St. slender, flexuose, naked below. L. lower small, loosely imbricate, upper much longer, ovate lanceolate, from a broad upright base declining, shortly hair-pointed ; per. l. three internal, erect, sheathing, longly pointed, basal cells rectangular, upper quadrate. Caps. ovate or ovate-globose, erect, smooth, exserted, lid acutely and obliquely rostrate, annulus broad. [Schp. Syn., p. 263. Dr. Braithwaite, l.c., 198.]

Dry quartzose rocks.   Spring.

Moncrieff Hill, Perth (Dr. Stirton); Dunkeld (Dr. B. White); Clova, in fr. (Fergusson); &c.

229. **G. montana,** *B. and S.* St. slender, dichotomous. L. erecto-patent, from an oval-oblong base, lanceolate, with a long hair-point, very concave, margin erect, basal cells diaphanous, quadrato-hexagonal, above minute, rounded, thickened, opaque. Caps. erect, on a short seta, ovate, small, brown, very smooth, lid obliquely rostrate. Calyptra large, long-beaked, annulus simple. Perist. teeth irregularly torn. [Dr. Braithwaite, l.c., 199. Sch. Syn., 264.]

Sandstone and granite rocks.   Spring.

Deeside, Aberdeenshire, 1869 (Prof. Barker and Mr. Roy); Bolt Head, Devonshire, in fruit (Mr. Holmes).

230. **G. Ungeri,** *Juratzka.* Compact, irregular, blackish-green, hoary tufts. St. short, simple or dichotomous. L. erecto-patulous, lower smaller, muticous, upper larger, lanceolate, from an obovate base, ending in a long, smooth hair-point, margin plane, basal cells quadrate, hyaline, above quadrate, then opaque and indistinct. Caps. small, oval, smooth, without annulus, exserted on an erect, pale brown seta, lid conical, obtuse. Calyptra cucullate. Monoicous. [Dr. Braithwaite, l.c., 198. Sch. Syn., 853.]

On earth in crevices of rocks at 1600 feet at Ballater (Rev. J. Fergusson).

c. Calyptra lobato-cucullate.

231. **G. elongata,** *Kaulfuss.* In loose, cushioned tufts, black below, innovations olive-green, with hoary tips. St. slender, repeatedly dichotomous, naked below, without radicles. L. patulous, lower lanceolate, muticous, upper elongate-lanceolate, obtuse, with the

apex diaphanous, margin erect, basal cells rectangular, hyaline at margin, becoming minute and quadrate above. Caps. ovate, erect, smooth, pale brown, on a straight seta, lid conical, obtuse, annulus narrow. Perist. teeth lanceolate, red, entire or slightly perforate. Calyptra multifid, long beaked. [Dr. Braithwaite, l.c., 199. Sch. Syn., 267.]

Alpine rocks. Summer.

Glen Callater and Glen Phee, Clova, 1868 (Fergusson); near Glasgow (Dr. Stirton).

232. **G. unicolor,** *Grev.* St. 1—2 inches, loosely cæspitose, naked below. Branches brittle, flexuose. L. erect, channelled, lanceolate-subulate, from an ovate base, obtuse, not hair-pointed, rigid, margin incurved, broadly nerved to apex. Caps. ovate, nearly erect, with a large annulus and a long, straight or slightly inclined beak. Dioicous.

Alpine rocks. Autumn.
Clova.

233. **G. atrata,** *Miel.* St. cæspitose, 1—2 inches. L. blackish, rigid, erecto-patent, lanceolate-subulate, carinate, margin reflexed, scarcely so obtuse as the last, with a thinner nerve, scarcely reaching to apex. Caps. ovate-oblong, on a longer seta, erect or subcernuous, annulus broad, lid rostrate, orange-coloured.

Alpine rocks. October—April.
Snowdon, Glen Callater, &c.

Section 5. *COSCINODON.*

Calyptra reaching to neck of capsule. Perist. teeth cribrose.

234. **G. cribrosa,** *Hedw.* Densely pulvinate, glaucous green, often black below. St. ½ inch, dichotomous.

L. erecto-patent, appressed when dry, from an ovate or oblong concave base, lanceolate, channelled, with a furrow on either side of the nerve below, rather obtuse, upper ones hair-pointed, margin erect, cells upper small, roundish, opaque, medial quadrate, basal next the nerve rectangular, contiguous, marginal more hyaline. Caps. subimmersed, erect, obovate, wide-mouthed, smooth, lid large, conical, with an erect, obtuse beak. Calyptra large, plicate, lobed at base. Perist. teeth widely lanceolate, beautifully cribrose, papillose at apex, reflexed when dry. Dioicous. [Rev. J. Fergusson in "Naturalist," v. 83, 1880. Sch. Syn., 287.]

Rocks and crevices.   April.
Coniston (Prof. Barker, April, 1867).

Section 6. *Rank uncertain.*

235. G. Stirtoni, *Schp.* In habit resembling *G. elongata* but differs thus:—L. scarcely squarrose-spreading, but patent and curved, upper continued into a long hair, margin distinctly thickened, upper cells minutely quadrate, lower lax-quadrate, thence at base shortly rectangular, scarcely sinuose nor elongate, hyaline at margin, thinly nerved, and blackish. [Sch. Syn., 270.]

Near Glasgow (Stirton, 1866).

40. **RACOMITRIUM,** *Brid.*

Sub-genus 1. *Campylodryptodon.*

Seta arcuate. Perist. teeth bifid almost to base.

· **236. R. patens,** *Brid.,* Bry. Univ., i. 192 (*Grimmia patens,* Bry. Brit., p. 158). In dark green or fuscous tufts. St. 2—4 inches, branched, decumbent, and naked below. L. spreading or slightly secund, elongate, lanceolate, gradually tapering to a blunt, slightly dentate apex, margin recurved below, nerve strong, two-winged at back; per. l. shorter. Caps. almost obovate, smooth, furrowed when dry, on a pale flexuose seta, mouth red, lid obliquely rostrate, annulus large. Calyptra five-lobed. Dioicous.

Moist alpine rocks. April, May.

Scotland, Wales, Ireland.

Sub-genus 2. *Dryptodon*.

L. solid, muticous. Calyptra with a rough, subulate beak. Seta erect, lid of capsule with an aciculate beak.

237. **R. ellipticum,** *Turner.* Blackish, rigid tufts. St. 1 inch, decumbent and naked below. L. spreading from an erect base, lanceolate, oblong, strongly nerved to apex, margins plane, thickened, upper cells sinuous, quadrate, lower longer, linear, sinuous. Caps. erect, ovate-globose, smooth, on a short, thick, twisted seta, lid large, conical, with a long, slender, subulate beak. Dioicous.

Moist alpine rocks. October—April.

Scotland, Wales, Ireland.

238. **R. aciculare,** *Linn.* St. 1—3 inches, cæspitose, decumbent and leafless, and slightly tomentose at base. Branches very leafy. L. spreading or secund, ovate-oblong or broadly lanceolate, obtuse, sometimes toothed at apex, to which the nerve does not reach, upper cells quadrate or ovate, lower linear, sinuous. Caps erect, oblong, smooth, with a small mouth, and on a longer and thinner seta, lid with a long, straight, subulate beak. Dioicous.

Wet mountainous rocks by streams.

November—April.

Var. β. DENTICULATUM. L. distinctly and distantly toothed at apex.

239. R. protensum, *A. Braun.* St. less rigid than last, and leafy at base. L. generally secund, from an oval, oblong base, lanceolate-subulate, obtuse, integrate, margin recurved, nerved to apex; per. l. sheathing. Caps. subcylindrical, thinner, on a pale seta, lid with a long, subulate beak. Calyptra five-lobed. Dioicous.

Moist alpine rocks.   April.

Wales, Derbyshire, Yorkshire, &c.

Sub-genus 3. *Racomitrium.*

Pl. nodose or fasciculate, branched. L. hair-pointed. Perist. teeth in two very long, filiform crura.

240. R. sudeticum, *Funk.* St. slender, decumbent and naked at base. L. spreading, recurved, keeled, lanceolate, tapering into a long, whitish, diaphanous, denticulate point. Caps. small, ovoid or elliptic, on a short seta, lid with a shorter, acute beak.

Alpine and subalpine wet rocks.   April.

241. R. heterostichum, *Hedw.* St. ¼—1 inch, base decumbent. Branches scarcely fasciculate. L. patent or falcate, secund, lanceolate, tapering to a long, white, denticulate point, margin recurved, areolæ long and sinuous below, above subquadrate. Caps. subcylindrical, mouth very small. Calyptra somewhat papillose at apex only, lid short, obliquely rostrate.

Rocks and walls.   March.

England, Wales, Ireland.

Var. β. ALOPECURUM. L. with short hair-points. Caps. smaller.

Var. γ. GRACILESCENS. L. obtuse. Caps. small, on a short seta.

242. **R. fasciculare,** *Schrad.* St. 1—2 inches, decumbent at base, with upright innovations, fasciculate. L. crowded, from a broadish, erect base, linear-lanceolate, spreading, muticous, margins recurved, areolæ long, narrow, sinuous. Caps. oval or oblong, with a long, subulate lid. Calyptra copiously papillose all over.

Wet rocks.   March.

243. **R. microcarpon,** *Hedw.* St. slender, fasciculate, with short branches. L. spreading, falcato-secund, lanceolate, tapering to a short point, areolæ all long and sinuous. Caps. small, ovate, thin and pale, with a robust, rostrate lid.

Highlands of Scotland.   Autumn.

244. **R. lanuginosum,** *Brid.* St. very long and slender, fasciculate. L. lanceolate, tapering into a long, strongly dentate point, sometimes secund or spreading, from an erect base, areolæ sinuous. Caps. small, ovoid, on a short, roughish seta, and with a long, straight, rostrate lid. Calyptra papillose above.

Mountains, walls, rocks, and heaths.   March—June.

245. **R. canescens,** *Hedw.* St. 2—4 inches, decumbent at base. L. erecto-patent or falcate-secund, ovate-lanceolate, tapering into a long, denticulate point, recurved from an erect base, areolæ sinuous. Caps. ovoid, eight-striate when dry, with a very long, subulate lid. Calyptra papillose above only.

Stony and sandy heaths.   March.

Var. *β.* PROLIXUM. Older innovations only with fasciculate ramuli.

Var. *γ.* ERICOIDES. Covered with fasciculate ramuli. L. squarrose.

Fam. 2. **Ptychomitriæ.**

41. GLYPHOMITRIUM, *Schwg.*

246. G. Daviesii, *Schwg.* St. ½ inch, cæspitose. L. linear- or ovate-lanceolate, spreading, entire, margin thickened and reflexed below, strongly nerved to apex, areolæ minute, larger and rectangular at base. Caps. erect, almost globose, with a reddish mouth and long, rostrate lid. Perist. teeth converging when moist, reflexed when dry. Calyptra large, laciniate at base.

Clefts of rocks. June, July.

Giant's Causeway, Llanberis; Glenarbuck, May, 1863 (W. Galt and McCartney); New Kilpatrick, Killin (McKinlay); Ardtun Mull (Dr. Black); Skye (Hunt); Blairlogie; Craigallion; Campsie Hills (Thompson and Galt), &c.

42. PTYCHOMITRIUM, *B. and S.*

247. P. polyphyllum, *B. and S.* St. 1—2 inches, tufted. L. linear-lanceolate, from a broadish, oblong base, spreading, nerved to the dentate, acute apex, margin reflexed below. Caps. elliptical, on a long, twisted seta. Calyptra plicate, lid long, subulate.

Mountainous rocks and walls. Winter.

Fam. 3. **Zygodonteæ.**

43. AMPHORIDIUM, *Schp.*

248. A. Lapponicum, *B. and S.* St. ½ inch, cæspitose, radiculose. L. spreading, linear-lanceolate, keeled, contorted when dry, nerve ceasing near apex, young ones bright green, older blackish-brown; per. l. ovate-lanceolate. Caps. scarcely exserted, turbinate, deeply eight-striate, urceolate when dry, lid with a short, oblique beak. Monoicous.

Crevices of Alpine rocks. Summer.
England, Scotland, Wales.

249. **A. Mougeotii**, *B. and S.* St. more than 1 inch, cæspitose, scarcely radiculose. L. fasciculate, recurved, narrowly linear-lanceolate, margin reflexed below, nerved to apex, scarcely contorted when dry. Caps. turbinate, urceolate when dry, 8-striate, lid with a long, very oblique beak. Dioicous.
Moist shady rocks. Summer.

44. ZYGODON, *Hook. and Tayl.*

a. Peristome absent.

250. **Z. viridissimus**, *Dicks.* St. ½—1 inch, fastigiate, radiculose below. L. squarrose, recurved, widely lanceolate, somewhat contorted when dry, minutely papillose, nerve sometimes slightly excurrent. Caps. obovate, obscurely 8-striate, lid with a long, oblique beak. Dioicous.
Trees. March, April.
Var. β. RUPESTRIS. (*Stirtoni*, Schp.). L. ligulate, not tapering to apex, nerve generally reddish and translucent, excurrent into a mucro.
Rocks. Not unfrequent.
England, Scotland, Ireland.

b. Peristome double.

1. Inner peristome fugacious.

251. **Z. conoideus**, *Dicks.* St. very short, cæspitose, fastigiate, densely radiculose. L. somewhat spreading, not recurved, linear-lanceolate, with plane margins, and nerved nearly to apex. Caps. arcuate, oval, tapering below, 8-striate, lid with a long, straight beak. Perist.

outer 8 short, recurved teeth, inner 8 yellowish cilia, often absent. Dioicous.

Trees.   May.

Scotland, Ireland, England.

2. Inner peristome persistent.

252. **Z. Forsteri,** *Dicks.* St. ¼—½ inch, densely tufted, with whitish radicles. L. erecto-patent, elliptic-lanceolate, nerved to apex or slightly excurrent, areolæ hexagonal (not, as in all the previous ones, dot-like), oblong and diaphanous at base. Caps. pyriform, striate, lid with a long, inclined beak. Perist. outer 8 acute, bigeminate teeth, inner 8 alternating cilia. Monoicous.

Trees. Rare.   May, June.

South of England.

253. **Z. Nowellii,** *Schp.* (*gracilis*, Wils. MS.). St. 1 inch or more, tufted, branched. L. patent, recurved, lower elliptic-lanceolate, upper from an oblong, concave base, narrowly lanceolate, twisted when dry, with plane margins, and denticulate near the apex, areolæ close and punctate above, large and pellucid below, nerve thin, hispid at back. Caps. oblong-cylindric, striate, lid rostrate.

Old walls.

Malham (J. Nowell), in fr., September, 1866; Littondale.

Fam. 4. **Orthotricheæ.**

45. **ULOTA,** *Mohr.*

a. *Repentes.*

St. creeping. L. scarcely crisped when dry.

254. **U. Drummondii,** *Greville.* L. linear-lanceolate, from a narrow, ovate base, blunt-pointed, scarcely crisped, margin hardly recurved. Caps. obovate-

clavate, on a long seta, 8-sulcate to base and contracted at mouth when dry. Calyptra with long, stiff
hairs. Perist. of 16 teeth in pairs, and no cilia.

Trunks of young trees. August.
Scotland, Ireland, England.

255. **U. Ludwigii,** *Brid.* L. subpatent or spreading,
narrowly linear-lanceolate, from an ovate base, slightly
contorted when dry, margin slightly undulate. Caps.
clavate-pyriform, striate only at summit, much contracted at mouth when dry. Perist. 16 teeth in pairs,
afterwards equidistant, occasionally with short cilia.
Calyptra densely pilose.

Trees. Rare. August, September.
England, Scotland, Ireland.

b. *Crispæ.*

St. erect. L. cirrato-crisped when dry.

256. **U. Bruchii,** *Hornsch.* St. short, tufted. L.
spreading, sharply linear-lanceolate, from an ovate,
concave base, subflexuose. Caps. pyriform, tapering
at base, on a long, twisted seta, with eight broad striæ,
and contracted at mouth when dry. Calyptra blackish-
yellow, with many furrows, deeply incised, and very
hairy. Perist. of 16 teeth in pairs, and 8, very seldom
16, cilia.

Trees. July, August.
Scotland, Yorkshire, Westmoreland, Sussex, &c.

257. **U. calvescens,** *Wils.* Differs from *Bruchii* by
its shorter leaves more narrowly reticulated. Calyptra
scarcely hairy. Caps. oval-oblong, with a long tapering
neck, lid more convex and shortly beaked. From
crispa by its longer seta and shorter capsule, not con-

tracted at mouth when dry, and by its smooth, glossy calyptra.

Trees.   June.

Killarney (Dr. Moore and Dr. Carrington); Dailly and Loch Doon (J. Shaw); Loch Lomond; Westmoreland.

258. **U. crispa,** *Hedw.* St. about 1 inch, tufted. L. linear-lanceolate, from an ovate, concave base, slightly waved at margin, very much contorted when dry. Caps. clavate, contracted at mouth when dry, eight broad, deep striæ, apophysis gradually tapering into the thick seta. Perist. 8 teeth, and 8, sometimes 16, cilia.

Trees.   July, August.

259. **U. intermedia,** *Sch.,* Syn. 305. Resembles *crispa.* L. longly linear-lanceolate, from an ovate or obovate base, cells of the middle base linear, subvermicular, the rest broader, rhomboid above, rectangular hyaline below, and orange-coloured. Caps. shorter, more narrowly striate, on a shorter seta, suburceolate when dry and empty, when old narrowly elongate, fusiform, scarcely contracted at mouth.

Trees.   June, July.

Not uncommon, frequently found with both *crispa* and *crispula.*

260. **U. crispula,** *Bruch.* St. very minute, tufted. L. small, linear-lanceolate, much crisped when dry. Caps. pyriform, small, thin, with 8 inconspicuous striæ, when dry short, truncate, urceolate, slightly contracted at mouth.

Trees. Not common.   May, June.

Yorkshire, Sussex, English lakes.

261. **U. phyllantha,** *Brid.* St. 1 inch, tufted. L.

linear-lanceolate, without a broad base, nerve extending to apex or exserted, where it is generally covered with gemmæ, crisped when dry. Fruit not known.

Rocks and trees. Generally near the sea.

c. *Strictifoliæ.*

262. **U. Hutchinsiæ,** *Sm.* St. about ½ inch, tufted. L. erecto-patent, imbricate and rigid when dry, broadly lanceolate, nerved to blunt apex, margin reflexed. Caps. clavate-pyriform, with eight broad striæ, slightly contracted at mouth when dry, apophysis tapering. Calyptra large, very hairy. Perist. 16 teeth in pairs, with 8 short cilia, sometimes wanting.

Mountainous rocks. Spring.

Wales, Ireland, Scotland, England.

46. ORTHOTRICHUM, *Hedw.*

a. *Anomalæ.*

Erect, dichotomous. L. solid, straight when dry, basal cells lax, hyaline. Caps. ovate or oblong, exserted.

1. Perist. simple, or the internal rudimentary.

263. **O. anomalum,** *Hedw.* Sp. Musc. (*non* Bry. Brit.). St. erect, simple, slightly branched. L. lower patulous, remote, upper erecto-patent, lanceolate from an ovate base, margin revolute, papillose, apical cells minute, rotundate-hexagonal. Caps. ovate-oblong, 16-striate. Calyptra brownish, hairy. Perist. t. pale, equidistant, erect when dry. [Bry. Eur., iii. t. 210. Schp. Syn., 308]

Trap rocks. Summer.

Aberdour, Fifeshire (Dr. Wood); Conway.

264. **O. saxatile,** *Brid.* (*O. anomalum,* Bry. Brit.,

p. 177.)   St. erect, sparingly branched.   L. narrow,
lanceolate, sharply acuminate, nerve thick, areolæ
narrow at base, crisped when dry.   Caps. narrow,
subcylindrical, much exserted, with 8 long striæ,
prominent when dry, and then with 8 shorter, alternating, spurious ones just below the mouth.   Perist.
teeth in pairs.   Calyptra furrowed, very hairy.   [Supp.
Bry. Eur., fasc. i. ii.]

Limestone walls and rocks.   Spring.

b. *Rupestræ.*

Taller.   L. minutely areolate above.   Calyptra more
or less hairy, campanulate.   Caps. striate, short-necked,
immersed or slightly exserted.   Perist. teeth 16, free.

1. Perist. simple.

265. **O. cupulatum,** *Hoffm.*   St. under 1 inch.   L.
spreading, straight and loosely imbricate when dry,
lanceolate, keeled, margin reflexed, densely papillose,
lower brownish, nerve distinct.   Caps. obovate, 16-
striate, urceolate when dry, with a shortly beaked lid.
Calyptra hairy.   Perist. simple, of 16 free equidistant
teeth, spreading when dry.

Rocks and walls.   May, June.

Var. β. NUDUM.   Calyptra naked, longer.   Caps.
oval-oblong, less striate.   Teeth erect.

266. **O. Sturmii,** *Hop. and Hornsch.*   In loose cushions.
St. short and erect, or longer and prostrate.   L. patent
and recurved when moist, incumbent when dry, margin
subrevolute, acutely costato-carinate, upper cells generally in two strata, papillose.   Caps. generally immersed,
obovate, with 8 obsolete striæ, when dry 8-ribbed and
constricted below the mouth.   Calyptra more or less

hairy, shining. Perist. teeth simple, 16 equidistant, erect when dry, slightly incurved. [Bry.Eur.,iii.t,109.]

Trap rocks. Summer.

Scotland, Ireland (Dr. Wood).

267. **O. Shawii**, *Wils. MS*. Resembles the last, but differs by its leaves being less solid and of looser texture at base, upper cells in one stratum always (not in two), by the fewer hairs on its shorter glossy white calyptra, and by its perist. teeth, densely papillose, white and reflexed when dry. [Supp. Bry. Eur., fasc. i. ii. Sch. Syn., 314.]

On an ash tree at Kilkerran Castle, Argyleshire, 1860 (J. Shaw). June.

2. Perist. double.

268. **O. rupestre**, *Schl.* St. 1 inch or more, cæspitose, creeping at base. L. broadly lanceolate, spreading, slightly recurved. Caps. pyriform, mouth large, scarcely exserted, indistinctly 8-striate. Calyptra large, yellow, with long hairs. Perist. 16 pale teeth in pairs (equidistant when dry), and 8 cilia.

Mountainous rocks. July, August.

c. *Obtusifoliæ*.

L. obtuse, margin incurved, imbricate when dry. Caps. immersed, striate.

269. **O. obtusifolium**, *Schrad.* In loose, yellowish-green tufts, brownish below. L. patulous, oblong, from an ovate base, apex obtuse, hyaline and minutely serrulate, concave, margin incurved, papillose at back; per. l. broader and less obtuse. Caps. oval, immersed, with 8 orange striæ. Calyptra long, naked, whitish, with a brown tip, lid convex, acuminate. Perist. teeth 8. bigeminate, reflexed when dry, alternating with 8 cilia. Dioicous. (Bry. Eur., iii. t. 208.)

On trunks of trees.   May.

York, Bristol, &c.

d. *Affiniæ*.

Caps. immersed or exserted, 8-striate, generally to base. Perist. teeth bigeminate, strongly reflexed when dry.

1. Internal perist. 8 or 16 cilia.

270. **O. affine,** *Schrad.*   St. ½—1 inch, tufted, branched. L. spreading, recurved, erecto-patent when dry, oblong-lanceolate, with a blunt point, margin revolute and slightly undulate, strongly papillose on both sides. Caps. elliptic-oblong, somewhat exserted, contracted when dry, striæ narrow. Perist. 8 pale teeth and 8 filiform cilia. Calyptra large, greenish-yellow, hairy, lid convex, margin yellow, beak pale.

Trees, walls, &c.   Common.   June, July.

271. **O. fastigiatum,** *Bruch.*   St. longer, tufted with fastigiate branches. L. broader, lanceolate, acuminate, gradually tapering to a point, suberect. Caps. almost pyriform, scarcely exserted, with broad striæ, lid broad, border reddish, long-beaked. Calyptra brown-ish-yellow, slightly hairy. Perist. teeth 8, and 16 broad, short cilia.

Solitary trees.   May, June.

Yorkshire, Sussex, &c.

272. **O. speciosum,** *Nees.*   St. 1 inch or more, tufted, branched. L. spreading, recurved, elongate, lanceo-late, somewhat pointed, papillose, margins recurved, nerve thin. Caps. shortly exserted, elliptic-oblong, faintly striate at summit, lid conical, beaked. Calyptra large, yellowish, with long hairs. Perist. 8 yellowish teeth, and 8 cilia.

Trees.   Rare.   July, August.

Montrose and Corrie Mulzie.

273. **O. patens,** *Bruch.* Tufts lax, yellow-green. L. spreading, recurved, from an ovate base, lanceolate, margin slightly revolute, very papillose. Caps. immersed, obovate, neck tapering, smooth, pale yellow, with 8 narrow striæ, truncate, and 8-furrowed when dry, lid with a short beak. Calyptra campanulate, slightly hairy, seldom naked. Perist. teeth bigeminate, pale yellow, cilia long, filiform. (Sch. Syn., 324.)

Trees. Rare. May.

Scotland.

274. **O. stramineum,** *Horns.* St. short, tufted. L. spreading, from an oblong base, narrowly lanceolate, acuminate, keeled, margin reflexed, papillose. Caps. ovate-pyriform, slightly exserted, striæ broad. Calyptra large, campanulate, purple-tipped, slightly hairy. Perist. 8 teeth, with 16 (sometimes only 8) cilia. Vaginula hairy.

Trees and rocks. June, July.

England, Scotland, Wales.

275. **O. pumilum,** *Swartz.* (*O. fallax,* Br., Wils., B. and S., but not Schp. Syn.). Minute, pulvinate. L. lanceolate, acute, carinate, margin revolute; per. l. longer, erect. Caps. oblong, with 8 orange striæ, neck gradually tapering into the seta. Calyptra long, shining, brown at apex. Perist. teeth 8, bigeminate, yellow, densely papillose, reflexed when dry.

Ash trees at Inverkip and Dailly, Ayrshire.

276. **O. fallax,** *Schp.* Syn. 327, not Bruch. (*O. pumilum,* Dicks, Bry. Brit., B. and S., Müller, &c.). Differs from the above in having a more oblong, thicker capsule, with deeper yellow striæ, with its neck shorter, abrupt, not gradually narrowed, and

with a shorter, more inflated calyptra. L. elliptic-lanceolate and obtuse.

On trees. Not common.   Spring.
England, Ireland.

277. **O. tenellum,** *Bruch.* St. ½ inch, tufted. L. spreading, broadly lanceolate below, above lanceolate-oblong or ligulate, obtuse, margin revolute. Caps. yellow-brown, exserted, subcylindrical, not contracted at mouth when dry, broadly and distinctly striate. Calyptra with a few short hairs, conico-campanulate, yellow. Perist. yellow, 8 teeth and 8 incurved cilia.

Trees.   May, June.
England, Ireland, Wales.

278. **O. pallens,** *Bruch.* St. erect, short, tufted. L. spreading, elongate, lanceolate or ligulate, obtuse, papillose, margins revolute. Caps. elliptic-oblong, with a large apophysis, scarcely exserted, slightly contracted at mouth when dry, striæ broad, yellow. Calyptra large, pale yellow, hairless. Perist. of 8 yellow teeth and 16 cilia.

Trees. Rare.   June.
Addingham, &c.

279. **O. diaphanum,** *Schrad.* St. scarcely ½ inch, tufted. L. spreading, ovate-lanceolate, tapering to a slender, diaphanous, serrulate point, margin recurved. Caps. subcylindric, almost immersed, faintly striate. Calyptra generally naked. Perist. 16 equidistant teeth, sometimes split at apex, and 16 cilia.

Walls, trees, and palings.   April.

e. *Pulchellæ.*

Tufts small, pulvinate. L. linear-lanceolate. Calyptra naked. Caps. small, exserted. Perist. double.

280. **0. pulchellum,** *Sm.* St. ¼ inch, tufted. L. spreading, soft, crisped when dry, linear-lanceolate, acuminate, margin recurved below, apical cells minute, papillose, basal lax, pale. Caps. pale, oval, with 8 reddish striæ. Perist. 16 reddish teeth in pairs, and 16 cilia. Calyptra campanulate, pale yellow, purplish at tip.

Trunks of trees. May.

f. *Lyelliana.*

Loosely pulvinate. L. long, squarrose. Caps. sub-immersed, substriate. Perist. double, teeth 16, free.

281. **0. Lyellii,** *Hook.* St. 2 inches or more, loosely tufted, with erect branches. L. much spreading, long, linear-lanceolate, wavy, scarcely serrate at apex, and studded with papillæ and brownish, gland-like bodies. Caps. elliptic-oblong, with a distinct tapering apophysis, and faint striæ, deeply sulcate when dry. Calyptra very large, brown tipped, with a few long, whitish hairs. Perist. 16 pale teeth, and 16 red, toothed cilia.

Old tree trunks. Rare in fr. July.

g. *Leiocarpa.*

Caps. immersed, without striæ, smooth. Perist. double, outer of 16 long, free teeth, revolute when dry.

282. **0. leiocarpum,** *B. and S.* St. 1—3 inches, tufted, branched. L. spreading, recurved, lanceolate, pointed, papillose, margin strongly revolute. Caps. large, pale brown, obovate, perfectly smooth and not contracted at mouth when dry, scarcely exserted. Calyptra hairy, sometimes naked. Perist. 16 teeth and 16 yellowish erose cilia.

Trees. April, May.

h. *Rivularia.*

Pl. flaccid, lurid green. L. broadish, soft. Calyptra naked. Caps. immersed. Perist. double.

283. **O. Sprucei,** *Mont.* St. ¼ inch, tufted. L. spreading, oblong-ovate or ligulate, apex rounded and tipped with an apiculus, scarcely reflexed, thinly nerved, not papillose. Caps. pyriform, contracted at mouth when dry, striæ broad, lid with a short beak. Calyptra campanulate, reddish tipped, large. Perist. 16 teeth, in pairs, yellowish, reflexed when dry, and 8, sometimes 16, cilia.

Trees near rivers.   May, June.

York, Matlock, Glasgow, &c.

284. **O. rivulare,** *Turn.* St. long, tufted, often floating. L. oblong-ovate, elongate above, flaccid, sometimes subsecund, obtuse, with a strong nerve and small papillæ, margin recurved below. Caps. shortly ovate, broadly and obscurely striate, almost immersed. Perist. 8 teeth in pairs, afterwards nearly equidistant, and 16 cilia. Calyptra large, campanulate, dull green, blackish at apex and base.

Rocks and tree-trunks at edges of streams.

June.

England, Wales, Ireland.

Tribe xiii. SCHISTOSTEGACEÆ.

47. **SCHISTOSTEGA,** *Mohr.*

285. **S. osmundacea,** *W. and M.* St. ¼—½ inch. L. bifarious, insertion vertical, lanceolate, pale green. Caps. small, subglobose, mouth large, lid convex, yellow, with a red border. Young plant, when growing in caves, emitting a beautiful golden-green light.

Sandstone caves and banks. Not rare.   March.

Tribe xiv. SPLACHNACEÆ.

Fam. 1. **Tayloriæ.**

a. Perist. absent.

48. **ŒDIPODIUM,** *Schw.*

286. **Œ. Griffithianum,** *Dicks.* St. ¼—½ inch, tufted, barren often much longer. L. obovate-spathulate, obtuse, fringed below, not nerved to apex. Caps. obovate or pyriform, neck tapering into a thick succulent seta, lid convex, obtuse. Calyptra membranous, hyaline below, brownish above.

Crevices of mountainous rocks. July, August.

b. Perist. of 16 teeth.

49. **DISSODON,** *Grev.*

287. **D. splachnoides,** *Schwg.* St. 1—4 inches, radiculose. L. erecto-patent, obovate-oblong or lingulate obtuse, margin plane, not nerved to apex. Caps. obovate, with a short, tapering neck, lid conical, pointed, columella exserted when dry.

Wet mountainous bogs. Autumn.
Scotland.

50. **TAYLORIA,** *Hooker.*

288. **T. tenuis,** *Dicks.* St. scarcely 1 inch. L. erecto-patent, broadly spathulate, shortly acuminate, serrate above, not nerved to apex, cells large, lax. Caps. oval, on a slender tapering seta, much contracted and wide-mouthed when dry, columella exserted.

Turfy soil on Scotch mountains. July, August.

Fam. 2. **Splachneæ.**

51. **TETRAPLODON,** *B. and S.*

289. **T. angustatus,** *L. fil.* St. ½—2 inches, tufted.

L. suberect, elongate-lanceolate, concave, narrowed
into a long, tapering, flexuose point, serrate, nerve
excurrent. Caps. ovate, on an obconical apophysis,
lid conical, obtuse.

Dung on mountains.   Rare.   July, August.

290. **T. mnioides,** *L. fil.* St. ½—3 inches, tufted.
L. suberect, obovate, or nearly elliptical, suddenly
narrowed into a long, piliferous, flexuose point, concave, entire, nerved to apex. Caps. elliptical, on a
large, obovate apophysis of about same width, lid
conical, obtuse.

Moist mountainous situations, on dung, &c.   Summer.

52. SPLACHNUM, *Linn.*

a. Caps. with an ovate or spherical apophysis.

291. **S. sphœricum,** *Linn. fil.* St. ½—1 inch. L.
roundish, obovate, acuminate, scarcely serrate, lower
smaller, nerved nearly to apex, apophysis not tapering,
roundish ovate, about same width as cylindrical capsule, lid mammillate.   Dioicous.

Dung in moist peaty places.   May, June.

292. **S. vasculosum,** *L.* St. ¼—1 inch, radiculose
L. roundish ovate, obtuse, or ovate acuminate, entire,
narrow at base, nerved nearly to apex, areolæ lax.
Caps. small, cylindrical, on a large globular apophysis,
lid convex.   Dioicous.

Elevated wet places.   July.

b. Apophysis pyriform.

293. **S. ampullaceum,** *L.* St. about 1 inch. L. lower
lanceolate, upper larger, obovate or oblong-lanceolate,
all serrate or sometimes entire, acuminate, nerved
nearly to apex, areolæ lax, seta dilated above into a

turbinate apophysis, bearing the small cylindrical caps., the whole shaped like the ancient *ampulla*, lid conical. Mono- or dioicous.

On dung in peaty places.   May, June.

Tribe xv. DISCELIACEÆ.
53. DISCELIUM, *Brid.*

294. **D. nudum,** *Dicks.* Stemless. L. few, ovate-lanceolate, entire, concave, round the base of the seta, generally buried, seta ½—1 inch. Caps. subglobose, reddish, cernuous, small, lid conical, acute.

Clay banks and beds.   February—April.
Near Manchester; Todmorden, &c.

Tribe xvi. FUMARIACEÆ.
Fam. 1. Ephemeræ.
54. EPHEMERUM, *Hampe.*
a. L. nerveless.

295. **E. serratum,** *Schreb.* (Ed. 1, *Phascum,* p. 26). Stemless, leaves lanceolate, erecto-patent, acuminate, serrated, connivent. Capsule large, subglobose, reddish, subsessile.

Sandy banks or fallows.   Spring or autumn.
Var. β. ANGUSTIFOLIUM. "Leaves narrower, linear-lanceolate, obscurely toothed. Caps. smaller."

296. **E. tenerum,** *Bruch.* Inconspicuous. L. broad, ovate-lanceolate, slightly denticulate at apex, very flaccid. Caps. small, pale yellow. Calyptra conical.

On the mud of dried-up pools.   Winter.
Weald of Sussex, Mr. Mitten.

b. L. nerved.

297. **E. cohærens,** *Hed.* Stemless, very minute. L.

lower ovate-lanceolate, almost nerveless, integrate,
upper elongate, lanceolate, keeled, erect, nerved to
apex, and serrated about half-way from summit. Caps.
subspherical, immersed, subsessile.

On the ground.   Winter.

298. **E. stenophyllum**, *Voit.* [*P. sessile*, B. and S.;
ed. 1, p. 27.] Very minute, almost stemless. L.
lanceolate-subulate, denticulate more than two-thirds
from summit, rigid, with an almost excurrent nerve.
Caps. sessile, small, rounded, brownish. Monoicous.

Clay and chalky heaths. Rare.   Autumn, winter.
Cheshire, Sussex.

Var. β. BREVIFOLIUM. L. shorter, linear-lanceolate,
slightly serrulate.

55. PHYSCOMITRELLA, *Schp.*

299. **P. patens**, *Hedw.* St. ⅛ inch. L. more or less
spreading, sometimes recurved, lower obovate-lanceolate, upper broadly obovate, spathulate, acuminate,
serrulate near the apex, concave, nerve ceasing below
apex. Caps. immersed, spherical, pointed, pale brown,
subsessile. Male fl. naked, in axil of a per. leaf.

Clay banks and fields.   Autumn.

Fam. 2. Funariæ.
56. PHYSCOMITRIUM, *Brid.*

300. **P. sphœricum**, *Schwg.* St. scarcely ¼ inch. L.
oval-oblong or slightly spathulate, acute, concave,
entire or obsoletely serrate, upper ones largest, nerved
nearly to apex. Caps. subglobose, mouth large, lid
large, conical. Calyptra lobed below.

Dried-up mud.   September—November.
Mere, Cheshire, 1834 (Wils.), J. Whitehead, October,
1870.

301. **P. pyriforme,** *Linn.* St. about ¼ inch, tufted. L. erecto-patent, lower distant, ovate-lanceolate, above spathulate, pointed, serrate, uppermost longer, erect, scarcely nerved to apex. Caps. globose-pyriform, erect, mouth small, lid conical. Calyptra subpersistent, toothed at base.

Moist banks and ditches. April.

57. ENTOSTHODON, *Schw.*

a. Perist. rudimentary.

302. **P. ericetorum,** *De Not.* St. ¼ inch. L. lower, distant, small, upper in a tuft, larger, obovate-lanceolate, acuminate, with a thickened, distantly serrate margin, not nerved to apex. Caps. small, subglobose, erect, lid almost flat.

Heaths, banks, stream sides, &c. March, April.

b. Perist. regularly 16-dentate.

303. **E. Templetoni,** *Schw.* St. about ¼ inch. L. lower distant, ovate-acuminate, upper in a rosaceous tuft, obovate, acuminate, nerved nearly to apex, scarcely serrulate. Caps. clavate-pyriform, upright, truncate when dry, neck tapering, lid plano-convex.

Crevices of rocks and shady places. July. England, Scotland, Ireland, Wales.

58. FUNARIA, *Schreb.*

a. Perist. imperfect.

304. **F. fascicularis,** *Dicks.* St. about ½ inch, tufted. L. patent, ovate-oblong or spathulate lanceolate, pointed serrate, not bordered. Caps. obovate or pyriform, subcernuous, tapering at base, on an upright seta, lid convex.

L

Fallow fields.   April.

b. Perist. perfect.

305. **F. calcarea,** *Wahl.* (*F. Muhlenbergii*, Schwg., et *Hibernica,* Hook., ed. 1, p. 126). Loosely cæspitose. L. lower remote, oblong-lanceolate, deflexed, upper erecto-patent, obovate-oblong, more or less acuminate, apiculate or flexuoso-subulate at apex, below which the yellowish nerve vanishes, margin obtusely serrate or subentire. Caps. turgid, brownish, on an upright (when dry twisted) seta, lid convexo-conical.

Walls, rocks, and stones. Not always confined to *limestone.*   Spring.

c. Perist. perfect. Caps. annulate, on a long, flexuose or arcuate seta.

306. **F. hygrometrica,** *Linn.* St. $\frac{1}{4}$—1 inch. L. lower scattered, upper ovate-lanceolate, concave, clustered into a bulb-like tuft, nerved to apex, margins reflexed. Caps. broadly pyriform, incurved, mouth oblique, with a corrugate border, deeply sulcate when dry, on a long, flexuose seta, lid convex, with a red border.

Banks and walls, old cinder-heaps, &c. Common.
                                        May—September.

Var. *β.* PATULA. St. slender, branched. L. undulate, spreading, twisted when dry.

Var. *γ.* CALVESCENS. Seta long and straight. Caps. slender, almost erect.

307. **F. microstoma,** *B. and S.* Habit of last, but smaller. L. in a comal bud. Caps. pyriform, turgid, scarcely furrowed when dry, on a thicker, arcuate seta, mouth very small, with a smooth border. Inner perist. very imperfect.

Damp stony places.   August, September.
Maresfield, Sussex (Mitten, 1864).

Tribe xvii.   BARTRAMIACEÆ.

Fam. 1.   **Amblyodonteæ.**

59.   AMBLYODON, *P. Beauv.*

308. **A. dealbatus,** *Dicks.*   St. ½—1 inch.   L. ovate-oblong and lingulate, lanceolate, acute, margins plane, slightly serrulate at apex, below which the strong nerve ceases.   Caps. clavate or pyriform, incurved, inclined, mouth oblique, lid conical, seta very long.

Wet mountainous places.   June, July.
England, Scotland, Ireland.

Fam. 2.   **Meesiæ.**

60.   MEESIA, *Hedw.*

309. **M. uliginosa,** *Hedw.*   St. ½—1 inch, radiculose. Branches fastigiate.   L. lanceolate or linear obtuse (upper longer), entire, margin recurved, scarcely nerved to apex.   Caps. pyriform, with a long, tapering neck, incurved, inclined, lid conical, truncate, seta very long.

Wet and boggy places.   July, August.

61.   PALUDELLA, *Ehr.*

310. **P. squarrosa,** *Linn.*   St. 2—6 inches, radiculose. L. obovate-lanceolate, pointed, recurved above the middle, squarrose, nerved to and serrulate at apex, margins recurved below.   Caps. elliptic-oblong, gibbous, with a short, thick neck, inclined, lid mammillate.

Boggy places.   No fruit found in Britain.   Summer.
Knutsford, Cheshire (Wilson, April, 1832); Terrington Carr, Yorkshire (Ibbotson, 1842).   Both places now drained, and the moss probably extinct.

62.   CATOSCOPIUM, *Brid.*

311. **C. nigritum,** *Hedw.*   St. 1 inch or more, densely radiculose.   L. erecto-patent, lanceolate, acute, margin

L 2

reflexed, entire, nerved nearly to apex, upper ones largest. Caps. small, globose, dark-coloured or black, cernuous, on a longish red seta, lid small, conical.

Moist alpine rocks, &c.   Spring.

Fam. 3.   Bartramiæ.

63.   BARTRAMIA, *Hedw.*

a. Caps. erect. Perist. single.

312. **B. stricta,** *Brid.* St. loosely tufted, glaucous, green. L. erecto-patent, rigid when dry, lanceolate-subulate, minutely serrate, nerve excurrent into a serrate arista. Caps. ovate-globose, furrowed when dry, seta four-sided at summit, twisted to right when dry, lid convex or mammillate. [Bry. Eur., iv. t. 316. Schp. Syn., 509.]

On the ground and stones.   Early summer.
Maresfield, Sussex (Geo. Davies), 1862.

b. Caps. cernuous, lid oblique. Perist. double.

313. **B. ithyphylla,** *Brid.* St. ½—2 inches. L. from an erect broad base, sharply bent back, and lanceolate-subulate, rigid, serrulate, not crisped when dry, broadly nerved to apex, nerve occupying all the subula. Caps. globose, almost erect or cernuous, deeply sulcate when dry. Synoicous.

Alpine and subalpine rocks.   June.

314. **B. pomiformis,** *Linn.* St. ½—2 inches. L. spreading, linear-lanceolate, not concave, doubly spinu-loso-serrate, rough, margin revolute below, crisp when dry, nerve excurrent into a spinulose arista. Caps. spherical, cernuous, on a flexuose seta, lid small, conical. Monoicous.

Dry sandy banks.   May.

Var. β. CRISPA. "L. longer, less crowded. Branches often longer than fruit-stalk."

315. **B. Halleriana,** *Hedw.*   St. 1—3 inches, radiculose below.   L. spreading or secund, linear-subulate from a broad, pale, erect, sheathing base, rough, serrate, nerve excurrent.   Caps. subglobose, on a short, curved seta.   Monoicous.

Moist alpine and subalpine rocks.   June, July.

316. **B. Oederi,** *Gunn.*   St. 1—3 inches.   L. linear-lanceolate, recurved from an erect, not sheathing base, crisped when dry, margins recurved, serrate above, keeled, smooth, nerve serrate at back, scarcely excurrent.   Caps. small, globose, oblique, lid convex.

Moist shady rocks.   May.

64.   CONOSTOMUM, *Swartz.*

317. **C. boreale,** *Swartz.*   St. $\frac{1}{2}$—2 inches, tufted, radiculose.   L. imbricate, lanceolate, acuminate, keeled, serrate, nerve excurrent into a mucro.   Caps. globose or obovate, gibbous, cernuous, deeply sulcate when dry, lid large, beaked.

Summits of Scotch mountains.   August, September.

65.   BARTRAMIDULA, *B. and S.*

318. **B. Wilsoni,** *B. and S.*   St. about $\frac{1}{4}$ inch, branched.   L. ovate-lanceolate, acuminate, somewhat secund, nerved nearly to or beyond apex, serrulate above.   Caps. globoso-pyriform, generally pendulous, not striate, lid convex or conical.

Turfy soil on mountains.   October.
Scotland, Wales, Ireland.

66.   PHILONOTIS, *Brid.*

a. Branches fasciculate.   Monoicous.   Male flower
gemmiform.

319. **P. rigida,** *Brid.*   St. $\frac{1}{4}$—$\frac{1}{2}$ inch.   Branches

erect or recurved. L. erecto-patent, straight, rigid, lanceolate, finely serrulate, nerve excurrent, shortly aristate. Caps. large, subspherical, furrowed when dry, and cernuous, lid conical, pointed, seta erect. Inner perist. sometimes imperfect.

Shady banks, mountains.   September, October.

320. **P. adpressa,** *Fergusson.* "Plant widely cæspitose, erect, 2—3 inches, either dull glaucous green or reddish. L. papillose, erect when moist, with one wide plica on each side of nerve, incurved towards apex, slightly twisted when dry, widely ovate from an amplexicaul base, not acuminate, apex either obtuse or cucullate, with a very slight mucro, or in the more slender forms rather acute, margin denticulate, slightly reflexed, nerve very thick, continuous. Areolæ small, ovoid above, shorter and wider towards the base." [G. E. Hunt. Mem. Lit. and Sci. Soc., Manchester, vol. v. 102, 1872.]

Glen Prossen, &c., Clova (Fergusson); Glas Mhcal, Perthshire, 2500 feet (Hunt).

321. **P. fontana,** *Brid.* St. 1—6 inches, with reddish-black radicles. L. ovate-acuminate, short and appressed or lanceolate secund, or spreading and longer (generally plicate at base), nerve almost excurrent; per. l. obtuse, nerveless. Caps. subglobose, large, furrowed when dry.

Springs and wet places.   June.

[Var. β. ALPINA. St. short. L. ovate-lanceolate, mucronate. Alps.]

Var. γ. FALCATA. L. falcato-secund, nerve thick.

Var. δ. CÆSPITOSA, *Wils.* Densely cæspitose, shorter. L. shorter lanceolate, subimbricate, or longer, lanceolate, more or less secund, not plicate, apical branches short.

Walton Swamp, near Warrington (Wilson); Scotland, &c.

Var. ε. COMPACTA (*firma*, Ferg.). Very compact, densely tomentose. St. filiform. · L. uniform, imbricate when dry, shortly lanceolate, from an ovate, concave base, nerve excurrent or vanishing in apex, margin obsoletely serrulate, slightly papillose.

Scotland.

322. **P. seriata,** *Mitt.*, Musc. Ind. Orient. "L. spirally arranged, from a suberect base, patent towards apex, ovate-acute, plicate, margin distinctly reflexed, areolæ linear above, small and ovoid towards base; per. l. from an erect, dilated base, widely spreading, cordate-triangular, obtuse, areolæ small, obscure, elongate-quadrangular, above large, linear, and reddish at base, nerve thick, indistinct, continuous or vanishing below apex, margin slightly denticulate." [Hunt, loc. cit., p. 103.]

Springs at head of Clova; Ben-na-Bourd (Gardiner).

323. **P. calcarea,** *B. and S.* St. about 2 inches. L. ovate-lanceolate, tapering gradually from middle upwards, concave, rigid, secund, margin serrulate, not reflexed, strongly nerved to apex, areolæ large, oblong, long hexagonal at base; per. l. acute, triangular, from a broad, erect base, nerved to apex. Caps. subglobose, inclined or cernuous.

Wet places. July.

67. BREUTELIA, *Schp.*

324. **B. arcuata,** *Dicks.* St. 1—6 inches, with reddish-brown radicles. L. ovate-lanceolate, from a broad, erect, sheathing base, scabrous, serrulate, striate, squarrose. Caps. subglobose, almost pendulous, on an arcuate seta, furrowed when dry.

Waterfalls and wet rocks.   September, October.

Tribe xviii.   Bryaceæ.

68.   MIELICHHOFERIA, *N. and H.*

325. **M. nitida,** *Funk.* "L. erecto-patent, larger and more crowded above, ovate-lanceolate, serrated at apex. Caps. suberect, pyriform, lid conical, very short."

Var. β. GRACILIS. More densely tufted. L. shorter, more crowded, imbricate. Caps. erect. [Wils. Bry. Brit., p. 263.]

Type not British; var. β. only found at head of Glen Callater, 1830 (Dr. Greville); again in same locality, 1868 (Fergusson and Roy); Ingleby, Yorkshire, 1862 (Mudd).

69.   ORTHODONTIUM, *Schw.*

326. **O. gracile,** *Wils.*   St. ½ inch, slender, tufted. L. long, linear, setaceous, carinate, flexuose, entire, faintly nerved nearly to apex. Caps. obovate-clavate, gradually tapering at base into the seta, inclined, lid long, beaked. Calyptra very small.

Sandstone rocks.   March.

Yorkshire and Cheshire. Not found elsewhere in Europe.

70.   LEPTOBRYUM, *Schp.*

327. **L. pyriforme,** *Linn.*   St. scarcely ½ inch. L. lower lanceolate, entire, upper linear-setaceous, flexuose, serrate at summit, nerve sometimes reaching apex. Caps. pyriform, pendulous, on a slender, flexuose seta, lid convex, mammillate.

Rocks. Frequent.   May, June.

71. WEBERA, *Hedw.*

Sub-genus 1. *Pohlia.*

Comal much longer than stem leaves. Caps. long-necked, narrowly pyriform, suberect or cernuous, internal peristome with the cilia very short or absent.

a. Monoicous. Male flower gemmiform, terminal.

328. **W. acuminata,** *Hoppe.* St. ½—1 inch, simple or branched. L. rigid, lower ovate-lanceolate, upper linear-lanceolate, larger, margins recurved below, nerved to serrulate apex, sometimes secund. Caps. almost horizontal, narrowly clavate, arcuate, tapering at base, lid sharply conical. (There are many varieties.) Crevices of rocks and mountainous districts.

August.

b. Antheridia hypogynous, axillary.

329. **W. polymorpha,** *H. and H.* St. ¼—½ inch, seldom branched. L. lower ovate-lanceolate, small, scattered, upper oblong, lanceolate, longer, crowded, apex in all serrate. Caps. oblong, pyriform, horizontal or drooping, with a short neck, slightly constricted at mouth when dry, lid mammillate. Perist. cilia almost absent.

Scotch and Welsh mountains, &c. Summer.

330. **W. elongata,** *Dicks.* St. ¼—1 inch, one inno-vation from floral apex. L. lower ovate-lanceolate, scattered, upper longer, linear-lanceolate, all serrate at apex, margin recurved to middle, apical cells linear, subvermiform, basal hexagono-rectangular, nerve ex-current into a subula. Caps. very long and slender, clavate or elliptic, with a long, distinct neck, inclined, upright when dry, lid convex, obliquely beaked. Inner perist. with cilia, sometimes rudimentary.

Rocks and walls in mountainous districts.   August.

Sub-genus 2.   *Webera.*

L. broader, texture lax, comal l. less elongate.   Caps.
short-necked, inclined and pendulous.   Internal peristome a broad membrane, with cilia.

a. Monoicous.   Antheridia in axil of comal leaves.

331. **W. nutans,** *Schreb.*   St. $\frac{1}{4}$—2 inches, with innovations.   L. spreading, with margins reflexed below,
lower ovate-lanceolate, entire, upper linear-lanceolate,
serrulate at apex, nerve thick, reddish, shining.   Caps.
oblong obovate, large-mouthed, with a short neck, lid
small, mammillate.   Perist. teeth red, pale at apex,
int. pale yellow, cilia as long as outer.

Sandy heaths, &c.   Spring.

b. Monoicous or dioicous.

332. **W. cruda,** *Schreb.*   St. 1—2 inches, cæspitose,
radiculose below.   L. lower ovate-lanceolate, with
plane margins and reddish nerve, upper gradually
passing upwards into linear-lanceolate, with serrate
apex.   Caps. oval-pyriform, from suberect to horizontal
or even pendulous, lid convex, apiculate.   Perist. teeth
pale, shining, cilia bi-ternate, perfect.   The leaves
are generally shining and transparent.   Antheridia
in monoicous plants intermixed with archegonia, in
dicious, subdiscoid axillary.

Mountainous banks and rocks.   July.

c. Dioicous.

333. **W. annotina,** *Hed.*   St. $\frac{1}{2}$—1 inch, cæspitose.
L. lower lanceolate, erecto-patent, entire, upper longer,
serrulate at apex, margins reflexed below, nerve excurrent.   Caps. narrow, pyriform, with a long, tapering

neck, inclined on a long, red seta, lid conical, pointed.
Barren fl. terminal. Inner perist. with cilia. Annulus
distinct, compound.

Sandy banks and quarries.   May, June.

334. **W. Ludwigii,** *Spreng.*   St. about 1 inch, decumbent and blackish below; st. l. remote, lower ovate,
muticous, passing upwards into ovate-lanceolate and
lanceolate, serrulate at apex, margins reflexed below,
nerve purple, vanishing below apex, cells lax, rhomboid.   Caps. obovate-pyriform, subpendulous on a
reddish, slender seta, 1 inch long, lid mammillate.
Perist. teeth large, yellow, internal membrane pale,
with bi-ternate cilia.

Scotch and Welsh mountains.   September.

Var. β. ELATA, *Schp.* St. 2—3 inches. L. ovate-
lanceolate, much acuminate, apex serrate, when dry
contracted, subflexuose.

[N.B.—This species seems to be identical with *W.
Breidleri,* Jur., of ed. 1, p. 187, and its var. is *W.
Schimperi,* Wils., on same page.]

335. **W. carnea,** *L.*   St. ¼ inch, cæspitose, reddish.
L. lower ovate-lanceolate, upper narrower, all serrate
at apex, and loosely reticulate, margin not reflexed,
nerve reddish.   Caps. ovate-oblong or shortly pyriform,
on a thick, succulent, reddish seta, sharply curved at
summit, lid large, convex, shortly pointed.   Annulus
none.   Perist. large, dark-coloured when dry, internal
membrane and cilia shining.

Moist clay banks.   April.

336. **W. Tozeri,** *Grev.* (*Epipterygium,* Lind.).   St.
¼—½ inch, gregarious.   L. lower remote, obovate,
narrow, reddish, upper crowded, longly acuminate,
apiculate, all bordered, entire, nerved (reddish) half-

way, areolæ lax, large.   Caps. obovate or pyriform,
pendulous, lid conical.

Shady banks.   Rare.   March, April.
Devonshire, Cornwall, Sussex, Ireland.

337. **W. albicans,** *Wahl.* (*Wahlenbergii*, Schw.).
Tufts soft, glaucous, green.   St. ½—1 inch, reddish,
cæspitose.   L. lower ovate-acuminate, upper elongate,
lanceolate, all serrate at apex, margins scarcely reflexed,
areolæ loose.   Caps. broadly pyriform, with a short
neck, and wide-mouthed when dry, subpendulous.
Annulus none or imperfect.   Perist. teeth large, inner
with cilia, lid small, conical.

Wet banks and rocks.   May.

72. ZIERIA, *Schp.*

338. **Z. julacea,** *Sch.* (*B. Zierii*, Dicks.).   Silvery-
reddish tufts.   St. ½—2 inches.   Branches julaceous.
L. roundish, ovate-acuminate, entire (comal oblong,
lanceolate), margins not recurved, not nerved to apex,
areolæ lax, chlorophyllose at base only.   Caps. large,
clavate-pyriform, gibbous, with a long, slender, tapering neck, incurved, cernuous, lid small, conical, acute.
Inner perist. longest, with imperfect cilia.

Crevices of mountainous rocks.   October, November.
England, Scotland, Ireland.

339. **Z. demissa,** *Hornsch.*   St. ¼ inch, tufted; st. l.
ovate-lanceolate, nerve vanishing, those of the coma
oblong, lanceolate, acuminate, margin recurved, nerve
excurrent; per. l. lanceolate, with longer points, areolæ
lax.   Caps. clavate-pyriform, much incurved, cernuous,
seta "curved above like a swan's neck."   Inner perist.
longest.

Rocks.   Rare.   August, September.
Breadalbane Mountains.

73. BRYUM, *Dill.*

Sub-genus 1. *Cladodium.*

Internal peristome composed of cilia and processes adherent to the teeth or free. Cilia imperfect or perfect, without appendages.

Synoicous or monoicous.

340. **B. pendulum,** *Hornsch.* (*cernuum,* Hedw.). St. ½—1½ inch, tufted, branched, very radiculose. L. ovate, acuminate, concave, rigid, nerve much excurrent, sometimes serrulate at apex, margins recurved. Caps. oblong-oval or pyriform, mouth small, neck not tapering, pendulous, lid small, conical, apiculate. Inner perist. adherent to outer, cilia and processes long, only partly free. Annulus large. Synoicous.

Walls and rocks. May, June.

341. **B. rufum,** *Ferg.* Tufts loose, dark brown. St. slender, short, with few radicles. L. enlarged toward top, forming a gemmiform tuft, brownish, twisted when dry, patent, then incurved from the middle, lower ovate-acute, margin scarcely recurved, nerved to or slightly beyond apex, upper ovate-lanceolate, carinato-concave, never red at base, margins entire, recurved, with a narrow border, nerve excurrent, into a smooth, yellowish point, cells mostly without chlorophyll, hexagonal above, rectangular at base. Caps. pendulous, at length looking upwards owing to curvature of summit of seta, obovate or oblong-pyriform, bladdery, not constricted below the very small mouth, which is not oblique, neck as long as capsule, lid conical, very small, obtusely apiculate. Annulus of 3 or 4 pale yellow cells. Perist. small, outer teeth lanceolate-subulate, pale yellow above, fuscous below,

inner about same length, adherent to outer.   Cilia not
seen.   Seta 4—7 lines long, reddish-brown.   Synoicous.
Rev. W. Fergusson, in "Naturalist," N.S., v. p. 82, 1880.

Loose earth and limestone rocks.   July.
Litton (Whitehead, 1879).

342. **B. inclinatum,** *Swartz.*   St. short, tufted,
branched.   L. ovate-lanceolate, entire, acuminate, margin slightly recurved, nerve reddish, much excurrent.
Caps. almost elliptical or pyriform, inclined or pendulous, on a long seta, with a small mouth, lid conical,
sharply pointed.   Perist. inner generally without cilia,
and free.   Synoicous.

Walls, banks, and decayed trees.   May.

343. **B. Warneum,** *Bland.*   St. about ¼ inch, tufted,
branched.   L. ovate-acuminate or oblong-lanceolate,
scarcely concave, serrate at apex, margins recurved
below, plane above, nerve excurrent, frequently spinose.
Caps. obovate, pyriform, pendulous, bordered at mouth
when dry, lid small, convex, mammillate.   Inner perist.
membrane adherent, processes free, cilia few.   Monoicous or synoicous.

Muddy places.   August, September.
Scotland, Southport, &c.

344. **B. lacustre,** *Brid.*   St. ¼ inch, cæspitose.
L. lower ovate-apiculate, upper ovate-lanceolate, all
entire, very concave, margins recurved, nerve reddish,
vanishing below the entire apex, or excurrent into a
short apiculus; per. l. narrower.   Caps. roundish,
pyriform, with a tapering neck, scarcely pendulous, lid
small, pointed.   Annulus present.   Inner perist. par-
tially adherent, cilia rudimentary.   Synoicous.

Moist sandy places.   May, June.
Yarmouth, Suffolk, &c.

345. **B. Barnesi,** *Wood.* St. short, branched. L. soft, patent, imbricate when dry, lower small, ovate, shortly acuminate, scarcely nerved to apex, upper 'ovate-lanceolate, much acuminate with the excurrent nerve, entire, margin plane or very slightly recurved ; upper l. with gemmæ in axils. Fruit not known.

Levens, Westmoreland (Barnes).

346. **B. Marrattii.** St. about ¼ inch, gregarious. L. broadly elliptical, blunt-pointed, entire, very concave, nerve vanishing below apex, marginal cells narrow ; per. l. narrower and longer. Caps. small, turbinate, almost globose, tapering at neck into the slender, flexuose seta, pendulous, lid small, conical, apiculate. Perist. outer red, inner imperfect, adhering in all its length to outer. Monoicous. September.

Southport, 1854. Tent's Muir, near Dundee.

347. **B. calophyllum,** *R. Br.* St. about ¼ inch, reddish, gregarious; st. l. broadly orbiculate-ovate, obtuse; comal l. ovate or oval oblong, shortly and obtusely acuminate, subsucculent, margin plane or slightly recurved, concave, entire, nerved almost to apex. Caps. oval-oblong, neck not tapering, pendulous, lid small, conical, slightly pointed. Perist. outer teeth brownish, yellow at apex, inner free, mostly without cilia. Monoicous. September.

Southport, with the last ; Ashton-under-Lyne.

348. **B. u'iginosum,** *B. and S.* St. ½—1 inch, cæspitose, branched. L. deep green, lower ovate-acuminate, upper ovate-oblong and elongate lanceolate, margin reflexed below, excurrent nerve toothed. Caps. elongate, pyriform, incurved, horizontal or inclined, tapering into the long, curved seta, mouth oblique, lid small, convex, pointed. Monoicous.

Wet places near streams.   August.

Sub-genus. *Bryum.*

Caps. inclined or pendulous, ovate or oblong-pyriform, rarely somewhat incurved. Perist. internal entirely free, basal membrane broad, processes long, perfect, ciliolæ of equal length.

a. Synoicous.

349. **B. intermedium,** *W. and M.* St. about ½ inch, tufted, branched. L. imbricate, lower ovate-lanceolate, upper oblong and elongate lanceolate, acuminate, margins recurved, nerve reddish, excurrent into a long, remotely toothed arista. Caps. pyriform, narrow, slightly incurved, subpendulous, tapering into a longish neck, lid conical, pointed. Inner perist. with cilia.

Walls and rocks.   Frequent.   June—December.

350. **B. bimum,** *Schreb.* St. ½—1 inch, tufted, sometimes branched, with purplish radicles. L. ovate-lanceolate, decurrent, semi-amplexicaul, shortly apiculate, occasionally serrate at apex, margins recurved, nerve brownish, shortly excurrent. Caps. oblong-pyriform, tapering at neck, pendulous, constricted at mouth when dry, lid large, convex, mammillate. Int. perist. with the basal membrane very broad.

Wet and boggy places.   June, July.

Var. *β.* CUSPIDATUM. L. with long, bushy points, margined.

Walls, &c.

351. **B. torquescens,** *B. and S.* St. ¼—1 inch, tufted, radiculose. L. lower ovate-lanceolate, upper oblong lanceolate, all shortly pointed, entire, margin recurved, slightly twisted when dry, nerve red, excurrent, into a smooth point. Caps. large, obconical or clavate, sub-

pendulous, neck tapering, lid conico-convex, acutely rostellate.

Rocks and walls. Rare. June, July.
Oxford (H. B.).

352. **B. provinciale,** *Phil.* (*B. Billarderii*, Schw.). St. ½—1 inch, branched, radiculose. L. crowded in tufts at top of branches and stems, ovate-lanceolate, and obovate at summit of stem, serrate and plane at apex, apiculate, margins recurved below, nerve reddish, shortly excurrent. Caps. pyriform, broadest below the middle, tapering at neck, pendulous, lid conical, pointed.

Hurstpierpoint, on old ant-hills. Barren. Summer.

b. Monoicous.

353. **B. pallescens,** *Schleich.* St. 1—2 in., branched, cæspitose, with purplish radicles below. L. lower remote, ovate, nerve vanishing, upper oblong acuminate or ovate-lanceolate, nerve slightly excurrent, margin reflexed, generally serrate at apex. Caps. clavate pyriform, pendulous, tapering, contracted at mouth when dry, lid convex, pointed. Inner perist. with cilia.

Rocks and walls. July, August.

[Var. β. BOREALE. Caps. smaller, suberect.

Var. γ. CONTEXTUM. St. long, much branched. Caps. ventricose, shorter, subpendulous.

Var. δ. SUBROTUNDUM. St. and l. smaller. Caps. almost globose, seta curved.]

354. **B. Sauteri,** *B. and S.* St. cæspitose, much branched. L. erecto-patent, ovate acuminate or oblong-lanceolate, elongated, very concave, margins entire, plane, nerve thick, excurrent into a mucro, wings at base with hyaline cells; per. l. narrower. Caps. clavate-pyriform, slightly incurved, pendulous, solid,

sanguineous, lid shortly conical.   Peristome as in last.

Teesdale (Spruce); Scotland (Mitten).   July.

c. Dioicous.

1. Male flower gemmiform.

355. **B. erythrocarpum,** *Schw.* (*B. sanguineum*, Ludwig, ed. 1). St. $\frac{1}{4}$ inch. L. distant, erecto-patent, ovate-lanceolate, elongate-lanceolate, pointed, generally serrulate at apex, margins scarcely recurved, nerve shortly excurrent into a mucro. Caps. oblong or obconico-pyriform, pendulous, blood-red when ripe, lid conico-convex, apiculate.

Heathy ground and walls.   June, July.

Var. β. RADICULOSUM. Caps. obconical, seta geniculate at base.

356. **B. murale,** *Wils.* St. $\frac{1}{4}$ inch, tufted, branched. L. of the coma erect, oblong lanceolate, margins reflexed, lower ovate-lanceolate, concave, shortly pointed, margins plane, loosely imbricate. Caps. oblong-pyriform, pendulous, deep purple or almost black when ripe, neck tapering, lid large, mammillate.

Mortar of old walls.   May, June.

Marple, Killarney, North Wales, Oxford, 1863 (H. Boswell), &c.

357. **B. atro-purpureum,** *W. and M.* (*B. erythrocarpon*, Brid., not Schw., ed. 1). St. $\frac{1}{4}$—$\frac{1}{2}$ inch, branched. L. erecto-patent, ovate acuminate, concave, margin reflexed below, entire, lower often reddish, remote, lanceolate. Caps. oval or oblong, neck not tapering, arcuate, pendulous, dark red or purplish when ripe, with a large mouth, lid convexo-conical, pointed, purplish.

Banks and walls.   May, June.

Var. β. GRACILENTUM, *Tayl.*

358. **B. alpinum**, *L.* St. ½—2 inches, cæspitose,
simple, radiculose at base. L. erecto-patent, imbricate,
lanceolate, margins recurved, serrulate at apex, nerve
purple, excurrent into a short mucro. Caps. oblong-
pyriform, pendulous, deep red, on a bent and arcuate
seta, lid mammillate. Whole plant reddish-purple and
shining.

Subalpine moist rocks. Fruit rare.   June.

Var. β. MERIDIONALE. L. very rigid, narrower, texture dense, upper cells linear-hexagonal, lower longer,
hexagono-rectangular.

359. **B. gemmiparum**, *De Not.* Densely tufted,
brownish-green, innovations variegated green. St.
¼—½ inch. L. imbricate when dry, lower ovate acuminate, upper ovate-elliptic and elliptic oblong, apical
short, narrow, all very concave, margin reflexed below,
nerve yellowish, ending in apex, cells rhomboid-hexagonal, lax and quadrate at angles, longer and narrower
at margin. Caps. ovate or oblong pyriform, inclined
or pendulous, lid very convex, shortly apiculate.
Perist. teeth yellow, processes pale, cilia slender.

Limestone, river banks, and damp places.

April, May.

River Usk, Breconshire (Rev. Aug. Ley, 1883).
Hitherto found only in the extreme south of Europe.
["Naturalist," N.S., viii. 185, 1883, H. Boswell.]

360. **B. cæspiticium**, *L.* St. ¼—1 inch, tufted,
branched, radiculose. L. ovate-lanceolate, pointed,
generally serrulate at apex, margin recurved but not
thickened, erect when dry, upper ones largest, nerve
much excurrent. Caps. oblong and elongate-pyriform,

pendulous or inclined, brown, slightly constricted when dry, lid large, mammillate, yellow.

Walls, rocks, roofs, &c.   May, June.

361. **B. argenteum,** *L.*   St. ¼—½ inch, tufted, in silvery-white patches. L. lower ovate, upper ovate-lanceolate, all entire except at apex, nerve not reaching pointed tapering apex, margins not recurved, areolæ very lax. Caps. oval-oblong, reddish-purple, pendulous, neck not tapering, lid faintly pointed.

Roofs, walls, ground, &c.   Winter, spring.

Var. β. MAJUS. St. longer. L. greenish, without points.

Var. γ. LANATUM. Smaller. L. with long points, without chlorophyll, silvery-white.

Largo, Scotland (C. Howie).

362. **B. capillare,** *Linn.*   St. ¼—1 inch, tufted. L. lower ovate-oblong, upper obovate-oblong, all with longish, slender points, concave, much contorted when dry, margin bordered with smaller cellules, reflexed, sometimes serrulate at apex, nerve reddish, vanishing or excurrent. Caps. subclavate or obovate, tapering, pendulous, or inclined on a long seta, only slightly constricted at mouth when dry, lid large, mammillate, red.

Walls, rocks, trees, &c.   Frequent.   June.

Var. β. MAJUS. St. longer. L. broader, and with a wider margin. Caps. larger and thicker. (Wilson says this is the most frequent English form.)

On walls.

Var. γ. MINUS. L. concave, imbricate. Caps. smaller.

Var. δ. FLACCIDUM. L. lower purplish, flaccid, not contorted when dry, distinctly serrate at apex.

363. **B. obconicum,** *Hornsch.*   St. short, tufted,

branched. L. oblong-ovate, elongate, much acuminate, very concave, entire, margin with a narrow border, recurved below, scarcely twisted when dry. Caps. obconical, pendulous, neck long, tapering, lid convex, apiculate.

Walls. June, July.

Barnard Castle, 1843 (Spruce), &c.

364. **B. Donianum**, *Grey.* St. short, branched. L. uppermost ovate-oblong, the rest elongate, acuminate, slightly pointed, not contorted, but slightly twisted when dry, margin thickened, serrulate at apex, nerve reddish below, brown above and excurrent. Caps. long, clavate, constricted at mouth when dry, sub-pendulous, lid mammillate.

Sandy banks and rocks. Rare. Summer.

Warrington (Wilson), Hurstpierpoint (Mitten), Winchelsea (Jenner), Penzance (Curnow), Scotland.

365. **B. pallens**, *Swartz.* St. ¼—1 inch, branched. L. reddish, lower remote, patulous and recurved, ovate-lanceolate, slightly decurrent, upper oblong and elongate-acuminate, margins thickened, recurved below, plane and entire above, nerve excurrent. Caps. oblong, pyriform, with a long, tapering neck, pendulous or horizontal, mouth small, but not contracted, lid small, convex, oblique, mammillate.

Near springs and ditches. June.

366. **B. barbatum**, *Wils. MS. (Stirtoni*, Schp.). St. about 1 inch, branched, slender, red, and copiously beset with reddish-brown radicles from base to summit. L. suberect, ovate, rather suddenly tapering into a longish, sparsely toothed subula, uppermost broader, more shortly pointed, spreading, all concave, strongly nerved, margins plane, not recurved, areolæ very lax

and transparent.   The only specimen I have bears no fruit.

Ben Ledi (Dr. Stirton).

367. **B. origanum**, *Boswell.*   St. elongate, 1 inch or more, copiously radiculose and forming dense, soft tufts. L. ovate and ovate-lanceolate, shortly pointed, scarcely acuminate, concave, nerved almost to apex, cells leptodermous, oblong and nearly rectangular, margins plane, slightly recurved when dry, formed of a single row of narrower cells.   Tufts dense, foliage full green, young leaves rosy-pink at summit, old and lower brown.   H. Boswell, in " Naturalist," N.S., v. 33, 1879.

Shady old wall, Teesdale, June, 1879 (J. S. Wesley).

2. Male flower discoid.

368. **B. Duvalii**, *Voigt.*   St. tufted, decumbent when old, elongate, branched.   L. patulous, remote, broadly ovate-lanceolate, decurrent, cirrhate when dry, scarcely nerved to apex ; per. l. inner lanceolate, erect.   Caps. obovate-oblong, regular, pendulous from a long, slender seta, contracted at mouth when dry, lid mammillate.   [Bry. Eur., iv. t. 371.   Sch. Syn., 458.]

Boggy places.   August, September.
Glen Lyon, Ben Lawers, Hartfell, Helvellyn.

369. **B. pseudo-triquetrum**, *Hedw.*   St. 1—3 inches, branched, erect, radiculose to summit.   L. lower ovate-lanceolate, upper narrower and longer, concave, all bordered, margins recurved, occasionally serrulate at apex, and slightly decurrent, nerve sometimes excurrent.   Caps. subcylindrical or elongate obconical, pendulous, lid small, mammillate.

Wet mountainous places and boggy ground.   July.

Var. δ. COMPACTUM. Tufts compact. Br. l. broader, flaccid. Caps. shorter, obovate.

370. **B. neodamense**, *Itzig.*, Regensb. Fl., 1841, i. p. 360. St. slender, cæspitose and tomentose, elongate, reddish and almost naked below, leafy above. L. lower roundish, oblong, obtuse, apex and margins involute, upper crowded, shortly oblong, inflated at base, margins revolute below, all cucullate. Caps. oblong-pyriform, pendulous, on a long seta, lid shortly apiculate.   Summer.

Southport Sands, where liable to inundation.

371. **B. turbinatum**, *Hedw.* St. ½—3 inches, sometimes branched. L. lower ovate-lanceolate, upper longer and narrower, concave, margins narrowly bordered, recurved, nerve rufous, excurrent in a mucro. Caps. globoso-pyriform, pendulous, when dry contracted at mouth, reddish-brown, lid convex, shortly apiculate.
June, July.

Ashton-under-Lyne, Fakenham, Norfolk, Oxford, &c.

372. [**B. Schleicheri**, *Schwg.* Plants with fastigiate, tumid branches, straw-green. L. lower small, remote, ovate, shortly acuminate, mucronate, concave, upper ovate-oblong, acuminate, reddish nerve excurrent into a mucro, or shortly cuspidate, apex denticulate, margin narrowly bordered, plane or slightly recurved. Caps. obconico-pyriform, constricted below mouth when dry, on a long seta.]

Var. γ. LATIFOLIUM. Broadly cæspitose, tumid, green. L. broadly ovate, rotundate-obtuse, or oblong-ovate, subacuminate, nerve excurrent, shortly mucronate, apex cucullate, margin erect, narrow bordered.

Variety only found in Britain in boggy places.

Ben More; Shetland (McKinlay).   August.

Sub-genus 3. *Rhodobryum.*

St. l. remote, subsquamiform ; comal l. in a rosulate, spreading tuft, acuminate, spathulate, serrate. Dioicous. Male flower discoid.

373. **B. roseum,** *Schreb.* St. 1—3 inches. L. lower small, scattered, lanceolate, upper in a large rosaceous tuft, spathulate, apiculate, serrate above, margin recurved, nerved nearly to apex. Caps. clavate-oblong or obovate, pendulous, lid mammillate.

Sandy shady banks.   September, December.

Sub-genus 4. *Anomobryum.*

Plants filiform. L. densely imbricate, julaceous, solid, shining, cells at base, hexagono-rhomboid and rectangular at apex, linear-vermiform.

374. **B. filiforme,** *Dicks.* (*B. julaceum,* Sm., ed. 1). St. 1—3 inches, tufted, with long, filiform branches. L. ovate or ovate-elliptical, obtuse, larger above, very concave, entire or minutely serrulate at apex, margin not recurved, not nerved to apex. Caps. oblong-obovate or pyriform, pendulous, lid small, convex, pointed.

Alpine and subalpine wet rocks. August, September.

375. **B. concinnatum,** *Spruce.* More attenuated than last, more regularly cæspitose, silky green. L. dense, imbricate, smaller, ovate or elliptic, more or less longly acuminate, muticous or shortly and acutely apiculate, entire or obsoletely denticulate at apex, nerved nearly or quite to apex, cells lax, narrowly hexagono-rhomboid, subflexuose, broader at base. Readily distinguished from *filiforme* by its narrower, less concave, acute leaves of laxer texture.

Near Kenmare (Dr. Taylor), Teesdale (Spruce), Ochills (Lyle).

4. Position uncertain.

376. **B. catenulatum,** *Schp.* Tufts taller, soft, slightly radiculose, incoherent. L. patent, curved, incurved imbricate when dry, giving the appearance of a thin chain (*catenulata*), ovate-lanceolate, decurrent, shortly acuminate, apex obsoletely serrate, somewhat concave, margin plane, scarcely bordered, nerve when young greenish-yellow, brown when old, vanishing below apex, cells subrhomboid, hexagonal, somewhat lax, chlorophyllose when young, empty when old. Fr. unknown. Ben Lomond (Stirton).

Tribe xviii. GEORGIACEÆ.

74. **TETRAPHIS,** *Hedw.*

377. **T. pellucida,** *Linn.* (*Georgia pellucida,* Br. M. Fl.). St. ½—1 inch. L. lower ovate-acuminate, nerved, reddish, upper larger, ovate-lanceolate, entire, nerve ceasing below apex, margins plane. St. bearing gemmiferous cups, l. of which are obcordate. Caps. elliptical, with a red border at mouth, on a long, reddish seta. Spring.

Decaying stumps and roots of trees. Common. Not often found in fruit.

75. **TETRODONTIUM,** *Schw.*

378. **T. Brownii,** *Dicks.* (*Georgia Brownii,* Br. M. Fl.). St. almost none, with long linear, radical leaves or ramuli; per. l. ovate-acuminate, entire, shortly and faintly nerved. Caps. oval-oblong, lid with an acute, oblong beak.

Sandstone rocks. Spring.

Tribe xx. MNIACEÆ.

Fam. 1. **Mniæ.**

76. **MNIUM,** *Dill.*

Section 1. *Branches stoloniferous. creeping or ascending from a subterraneous base.*

a. Margin thickened, simply serrate.

379. **M. cuspidatum,** *Hedw.* St. ½—1 inch, tufted, erect, radiculose, barren shoots procumbent. L. lower ovate or obovate, scattered, upper ovate-lanceolate, longer and narrower, all acuminate, with simply serrate margins, nerved nearly or quite to apex. Caps. ovate, inclined or pendulous, lid convex, obtuse. Synoicous.

Shady rocks and walls.   March, April.

380. **M. affine,** *Bland.* St. 1—3 inches, simple, erect, radiculose, barren shoots procumbent. L. lower oval-lanceolate, decurrent, scattered, upper much larger, crowded, oblong-elliptic, pointed, border narrow, simply spinuloso-serrate, nerved nearly or quite to apex, those of barren stems roundish, two-ranked. Caps. ovate-oblong, pendulous, lid convex, pointed. Dioicous.

Shady woods, banks, walls.   April, May.

Var. *β.* ELATUM. Stem and seta longer. Marshy places.

Var. *γ.* RUGICUM. Sterile shoots, erect, shorter. L. shorter, concave, obtusely serrate.

381. **M. undulatum,** *Hedw.* St. 1—3 inches, decumbent at base, sometimes branched, stolons with large leaves, and pendulous at apex. L. oval-oblong or ligulate, upper very long, all undulate, decurrent, and simply serrate, nerved generally to apex. Caps. generally several together, oval or oblong, pendulous, lid convex, pointed. Dioicous.

Moist shady banks and woods. April, May.

382. **M. rostratum,** *Schrad.* St. ½—1 inch, decumbent at base (barren, long, erect or creeping). L. lower ovate, upper oval-oblong, obtuse, in a terminal, spreading tuft, all simply and bluntly serrate, undulate, nerve slightly excurrent into a mucro. Caps. oval, inclined or pendulous, lid with a long, curved beak. Synoicous.

Moist shady rocks, &c. Common. April.

Section 2. *Basilar branches erect.*

a. Leaves with a thickened border, doubly dentate.

* *Lid mammillate.*

383. **M. hornum,** *L.* St. 1—3 inches, simple. L. linear-lanceolate, acuminate, rigid, very slightly decurrent, doubly spinuloso-serrate, nerve also spinulose, not reaching apex. Caps. large, oblong-ovate, cernuous, lid convex, mammillate, seta curved at summit. Dioicous.

Shady moist banks and woods. Common. May.

** *Lid conical, rostrate.*

384. **M. serratum,** *Schrad.* St. ½—1 inch, purplish, erect. L. lower reddish on nerve and border, all varying from obovate-lanceolate to spathulato-lanceolate, acuminate, much decurrent, doubly spinuloso-serrate, nerve vanishing in apex; per. l. lanceolate. Caps. ovate or oval, cernuous, lid large, conical, with a short beak. Synoicous.

Moist shady rocks and banks. May, June.

385. **M. orthorhynchum,** *Brid.* St. ¼—1 inch, simple. L. lower scattered, ovate-acuminate, decurrent, upper ovate-lanceolate, doubly spinuloso-serrate, nerve spinulose on back, all undulate and crisped when dry.

Caps. oval or subpyriform, horizontal, lid conical, with a blunt beak. Dioicous.

Woods, shady banks, &c.   Summer.
Yorkshire, Sussex, &c.

386. **M. riparium,** *Mitt. MS.* St. ½—1 inch, reddish below. L. orbiculate or broadly elliptical, much decurrent, apex rounded and tipped with a mucro, lower obscurely bordered, upper strongly so, doubly spinulosodentate, nerve thin but distinct, reddish, reaching apex, where it is spinulose at back, areolæ large, hexagonal, chlorophyllose. Caps., &c., as in *serratum.*

Watery places.   June.
Sussex (Mitten).

387. **M. spinosum,** *Voigt.* St. robust, subligneous. Branches flagelliform, subarcuate. L. lower small, squamiform, obtuse, margins plane, upper obovate or oblong, acuminato-spathulate, larger, crisped when dry, serration bigeminate, nerve reddish, excurrent. Caps. oval-oblong, reddish-brown, slightly inclined or horizontal, lid conical, with an obtuse beak. Dioicous.

                                          Summer.
Roots of trees, and shady subalpine rocks. Rare.
Ben Lawers (McKinlay).

b. Leaves scarcely bordered, serrate or entire.

* Lid convexo-conical, with or without apiculus.

388. **M. stellare,** *Hedw.* St. ½—2 inches, erect. L. oval-acuminate or ovate-lanceolate, simply serrate, decurrent, scarcely nerved to apex. Caps. solitary, ovate, horizontal or cernuous, lid convex, scarcely apiculate. Dioicous. (Does not fruit with us.)

Shady rocks and banks.   May, June.
Yorkshire, Surrey, &c.

389. **M. cinclidioides,** *Blytt.* St. 2—4 inches, sometimes with slender branches. L. lower oval, obtuse, scarcely pointed, upper large, oval, ligulate, obtuse (marginal cells narrower), almost entire, slightly undulate, nerved nearly to apex. Caps. oval, pendulous, lid convex, apiculate. Dioicous.

Wet and boggy places on mountains. Summer. Clova, &c.

c. Leaves with a border, margin entire.
Lid rostrate.

390. **M. punctatum,** *Hed.* St. ½—3 inches, erect, radiculose. L. obovate or roundish, obtuse, obscurely pointed, upper in a somewhat rosaceous tuft, thickened margin generally reddish, generally nerved to apex. Caps. oval, pendulous, lid conical, with a longish beak. Dioicous.

Wet shady places. February, March.

391. **M. subglobosum,** *B. and S.* St. 1 inch or more, erect. L. obovate or roundish, obtuse, not nerved to apex, border narrow, subcartilaginous, not coloured. Caps. roundish, contracted at mouth, subpendulous, lid small, conical, beaked. Synoicous.

Marshes and bogs. March.

77. CINCLIDIUM, *Swartz.*

392. **C. stygium,** *Sw.* St. 1—4 inches, tufted, erect, with purplish radicles. L. roundish, obovate, obtuse, pointed, rigid, very narrow at base, border cartilaginous, nerved to or beyond apex, nerve and border generally reddish. Caps. obovate or pyriform, pendulous, on a long seta, lid convex, obtuse, sometimes pointed. Synoicous.

Bogs.   June, July.
Malham Tarn (Nowell, Wilson), Suffolk (Rev. W.
Bloomfield), High Force, Teesdale (W. West).

Fam. 2. Aulacomniæ.

78. AULACOMNION, *Schwg.*

393. **A. androgynum,** *Linn.* St. less than 1 inch,
nearly simple. L. lower lanceolate, upper longer, all
denticulate at apex, not flexuose nor crisped when dry,
papillose, margin recurved. Caps. almost cylindrical,
gibbous, inclined, furrowed, lid short, conical, pseudopodia numerous, barren fl. gemmiform.
Dry shady woods and banks.   May, June.

394. **A. palustre,** *Linn.* St. 2—4 inches, branched,
beset with reddish radicles. L. oblong-lanceolate,
denticulate at apex, flexuose, undulate, margin reflexed,
crisped when dry, papillose on both sides, pseudopodia
less numerous. Caps. ovate-oblong, gibbous, cernuous,
curved, lid conical, with a blunt beak, barren fl. discoid.
Dioicous.
Turfy bogs and marshes.   May, June.

395. **A. turgidum,** *Wahl.* St. about 1 inch, scarcely
radiculose. L. turgid, imbricate, ovate-oblong, elongate, obtuse, concave, margin entire, reflexed, less
papillose, nerve vanishing in apex. Caps. ovate-oblong,
turgid, lid convexo-conical, shortly apiculate, barren
fl. discoid, pseudopodia absent.   Summer.
Whernside, Yorkshire (West and Lees); Perth.

Fam. 3. Timmiæ.

79. TIMMIA, *Hedw.*

396. **T. austriaca,** *Hedw.* St. 2 — 3 inches. L. linear-
lanceolate from a reddish-brown, sheathing, dilated

base, margins strongly serrate. Caps. ovate, pyriform, inclined, striate when dry, on a seta 2 inches long, lid rounded, mammillate. Perist. teeth inner smooth, entire.

Rocks. Very rare. Summer.
Forfarshire.

397. **T. norvegica**, *Zett.* St. radiculose. L. soft, slightly sheathing, lower elongate-elliptical, acute, acuminate, upper longer, narrowly linear, very concave, more or less denticulate, particularly near apex, nerve broad, upper cells papillose in front, lower behind.

Rocks.
Ben Lawers (McKinlay), &c.

Tribe xxi. BUXBAUMIACEÆ.

80. BUXBAUMIA, *Haller.*

398. **B. aphylla**, *Hall.* "St. almost none, buried. L. lower roundish, deeply toothed, upper fringed with long, ciliary processes. Caps. plano-convex, roundish, ovate, reddish. Outer perist. irregularly subdivided, thick and cellular." [Wilson.]

Rare. May.
Scotland, Yorkshire, &c.

399. **B. indusiata**, *Brid.* "Resembling the last, but caps. more erect, not flattened on the upper surface, of uniform texture and yellowish-green colour, covered with a soft membrane, which ruptures on the upper surface, the margins rolling back, somewhat like the indusium of a fern. Annulus narrow." [Dr. Braithwaite, Journ. Bot., viii. 226.] June.

On ground and rotten trunks. Chiefly in pine woods.
Near Ballater, 1847 (Cruikshank); Craigendinnie Hill, Aboyne, 1867 (Dickie and Roy).

81.   DIPHYSCIUM, *W. and M.*

400. **D. foliosum,** *W. and M.* St. almost none. L. long, narrow, linear, flexuose, with an obscure nerve, margin plane, sometimes toothed near apex; per. l. with a pale, thin blade, nerve excurrent into a long, rough bristle, and the innermost divided at apex into long, jointed cilia. Caps. immersed, ovate, oblique, gibbous, lid conical, pointed. Perist. teeth white.

Shady mountainous rocks.   August.

Tribe xxii.   POLYTRICHACEÆ.

82.   OLIGOTRICHUM, *De Cand.*

401. **O. hercynicum.** *Ehr.* (*incurvum*, Huds., Br. M. Fl.). St. ½—1 inch. L. rigid, erecto-patent, lanceolate, sheathing, margins inflexed, remotely serrate, lamellæ of nerve undulate, and spinulose at back. Caps. erect, cylindrical, plicate and oblique when dry, lid conical, pointed.

Alpine and subalpine barren soil.   June, July. Scotland; Todmorden (Nowell).

83.   ATRICHUM, *P. Beauv.*
[*Catharinea*, Ehr., Br. M. Fl.]

a. Monoicous.

402. **A. undulatum,** *Linn.* St. 1—2 inches. L. ligulate, margin undulate, thickened, with bi-cuspid, spinulose teeth, which also occur on back near apex, where lamellate nerve ceases. Caps. cylindrical, curved, lid with a long, curved beak.

Grassy places. Common.   October, November.

Var. *β.* MINOR, *Hedw.* L. crowded, shorter, less undulate. Caps. suberect, ovate-oblong, unequal, on a shorter seta. [Br. M. Fl., p. 40.]

Bare stony places. Not common.

Between Ben Lawers and Killin, 1865 (McKinlay) ; Islay, 1883 (Rev. A. Ley).

b. Dioicous.

403. **A. angustatum,** *Brid.* St. shorter. L. narrower, densely reticulate, margin serrate at apex *only*, less hispid beneath, with numerous lamellæ on nerve. Caps. suberect, cylindrical, straight or curved, lid dark purple, shortly rostrate. [Schp. Syn., p. 529.]

Bare sandy ground, heaths, &c. Winter.

Braes of Doune, fruit (McKinlay) ; Sussex—male (Mitten).

[*A. tenellum,* Röhl. The plants referred to this species as British belong to *undulata* or its var. *minor.*]

404. **A. crispum,** *James* (*laxifolium,* Wils. MS.). St. simple, slender. L. lower small, somewhat spathulate, upper much larger, erecto-patent, curved and flexuose when dry, oblong-lanceolate, slightly undulate, border thickened, toothed, nerve reaching apex, narrowly 1−4 lamellate, areolæ larger, hexagonal or rounded. Caps. obovate-oblong, suberect or inclined, wide-mouthed, teeth very short, irregular, seta red.

Lancashire; Oakmere, Cheshire; Saddleworth, Marsden (Whitehead), Hebden Bridge (Nowell), Yorkshire; Dartmoor. Male plant only found in Britain.

84. POGONATUM, *P. Beauv.*

Aloidea.—Pl. low, simple, gregarious. L. broad, involute (resembling those of an Aloe).

405. **P. nanum,** *Dill.* (*Polytrichum subrotundum,* Huds., Br. M. Fl.). St. short, not branched at apex. L. rigid, spreading, lanceolate-obtuse, sheathing, serru-

late above. Caps. almost globose, erect or inclined,
lid with a curved or oblique beak, columella not winged,
cylindric.

Moist shady sandy banks.   Spring.

Var. β. LONGISETUM. L. long, linear-lanceolate, seta
very long. Caps. oval-oblong.

406. **P. aloides,** *Hedw.* (*Polytrichum*, Br. M. Fl.).
St. less than 1 inch, branched at apex. L. rigid,
spreading, lanceolate, sheathing, serrate on margin
and back. Caps. erect, somewhat ovate-urceolate or
cylindrical, lid conical, beaked, columella with four
wings. Hairs of calyptra whitish.

Moist banks.   Spring.

Var. β. MINUS. Smaller generally. "Hairs of calyptra confluent below capsule."

Urnigera.—St. taller, dendroid.

407. **P. urnigerum,** *Linn.* St. 1—4 inches, branched
above, reddish below. L. spreading, linear-lanceolate,
from a short, broader, sheathing base, acute, serrate.
Caps. erect, narrowly cylindrical, regular, papillose, lid
convex, with a short, upright beak.   October, March.

Mountainous banks and sides of streams.

Var. β. HUMILE. St. short, simple. L. shorter,
straight. Caps. narrower, ovate, subcernuous.

408. **P. alpinum,** *Linn.* St. much longer, branched
above, decumbent below. L. linear-lanceolate, from a
long sheathing base, gradually tapering, serrate, back
spinulose, margin inflexed. Caps. generally inclined,
ovoid, tumid, smooth, lid small, with a long, curved,
oblique beak.

Subalpine localities.   June.

England, Scotland, Ireland.

Var. δ. CAMPANULATUM. St. short. L. narrower and shorter. Caps. apophysate. Calyptra campanulate.

85. POLYTRICHUM, *Brid.*

A. Caps. 6-angled. Apophysis indistinct.

409. **P. sexangulare,** *Floerke.* Barren st. 2—6 inches, fertile shorter. L. short, incurved, rigid, linear-lanceo-late from a broader base, margins plane, incurved, almost cucullate at serrulate apex. Caps. short, at first upright, afterwards cernuous, sometimes 6, some-times only 4 or 5-angled, lid rostrate. Calyptra short, and shortly villous.

Summits of Scotch mountains. August, September. Ben Lawers, Cairngorm, &c.

410. **P. gracile,** *Dicks.* St. about ½ inch, densely tufted. L. short, lanceolate, from a broad, sheathing base, wings pellucid, serrate, with prominent lamellæ on upper surface. Caps. ovate, obscurely 6-angled, lid conico-rostrate, beak oblique. Calyptra small.

Turbaries, &c. Frequent. Spring.

411. **P. formosum,** *Hedw.* (*attenuatum*, Menz., Br. M. Fl.). St. 3—6 inches, loosely tufted. L. spreading, linear-lanceolate, from a broad, glossy, sheathing base, acute, serrate, lamellate. Caps. 4, 5, 6-angled, fawn-coloured, lid long, conical, beak acuminate. Calyptra large.

Woods. June.

B. Caps. 4-angled. Apophysis very distinct, discoid.

412. **P. piliferum,** *Schreb.* St. about 1 inch, simple. L. lower short, appressed, upper much longer, from an ovate, sheathing base, lanceolate, prolonged into roughish hair-points, margins entire, inflexed. Caps.

faintly angular, almost ovate, erect, on a short seta.
Calyptra long, reddish.

Dry heaths.   May, June.

413. **P. juniperum,** *Hedw.* St. 1—6 inches. L. reflexed, spreading, linear-lanceolate, almost bristle-
pointed, margin entire, except a few teeth at apex,
much inflexed from middle upwards, somewhat spinulose at back, nerve excurrent into a short, reddish arista.
Caps. 4-angled, reddish-orange, on a seta 2 inches or
more long. Calyptra large, very hairy.

Heaths.   May, June.

414. **P. strictum,** *Banks* (*P. juniperum, β. strictum,*
Bry. Brit.). St. more slender, densely matted, branched,
closely interwoven with whitish tomentum. L. straight,
erecto-patent, imbricate when dry. shorter and narrower, pale glaucous green. Caps. shorter, cuboid,
acutely angled, rufous orange. Calyptra brownish or
white.

Mountain moors.   Common.   Summer.

415. **P. commune,** *L.* St. 6 or 8 inches, simple. L.
spreading, reflexed, linear-lanceolate, sheathing, whitish
and membranous at base, serrate on margin and back,
nerve broad ; per. l. with long, wavy hair-points.
Caps. short, upright, acutely 4-angled, afterwards
cernuous, on a very long seta, lid variable in size.
Calyptra large, reddish, very hairy.

Tufty and marshy places.   June.

Var. β. PERIGONIALE. Smaller. Inner per. l. longer;
l. nearly smooth at back.

Dry heaths.

Var. γ. MINUS. St. short. L shorter, less spreading ; per. l. less striking. Caps. much smaller, less
acutely angled.

Wet heaths.

Var. γ, β. FASTIGIATUM. St. branched, fastigiate. L. as in var. γ.

Very dry heaths.

Section II. *CLADOCARPI.*

Tribe xxiii. FONTINALACEÆ.

Fam. 1. Cinclidotæ.

86. **CINCLIDOTUS**, *P. Beauv.*

416. **C. fontinaloides,** *Hedw.* St. 2—5 inches, in long, straggling tufts, generally floating. L. crowded, spreading, flexuose, lanceolate, acute, with a thickened margin and strong, excurrent nerve; per. l. larger, sheathing, ovate-lanceolate, thinner. Caps. immersed, with a conical, beaked lid. Calyptra persistent, thick, split on one side.

Stones in rivulets, &c. March, April.

Fam. 2. Fontinalidæ.

87. **FONTINALIS,** *Dill.*

417. **F. antipyretica,** *L.* St. very long, often 1 foot, with long, spreading branches. L. ovate-lanceolate, very concave, keeled, nerveless, all on each branch with one margin reflexed on the same side, the other plane, sometimes serrulate near apex, olive or almost black when old. Caps. oval or ovate, immersed, lid long; conical, acute.

Streams and stagnant water. June, July.

Var. β. GIGANTEA, *Sull.* More robust, with fewer branches, pale green when young, rufescent or orange-coloured when old. Caps. smaller. Perist. teeth shorter, paler.

In slow flowing streams.

Var. γ. GRACILIS, *Lindb.* More slender. L. smaller, narrower, acute, less acutely keeled, cells narrower, often bifid on the keel, margin less reflexed.

Colder streams.

Malham Cove; Scotland.

418. **F. squamosa,** *L.* St. shorter, but elongate. Br. numerous, crowded, fasciculate, not spreading. L. lanceolate or oblong-lanceolate, concave, not keeled, margin not reflexed, nerveless, entire. Caps. similar to last.

Mountain rivulets.   June, July.

Fam. 3.   **Dichelymeæ.**

88.   **DICHELYMA,** *Myrin.*

419. [**D. capillaceum,** *Dicks.*] St. 3—6 inches, slender, brittle, with a few distichous, spreading branches. L. erecto-patent, secund, subulate-setaceous, keeled, with a long, excurrent nerve; per. l. long, convolute, nerveless. Caps. short, oval, almost immersed, lid large, conical, beaked.

Alpine rivulets.   Summer.
A doubtful native.

Tribe xxiv.   CRYPHEACEÆ.

Fam. 1.   **Hedwigiæ.**

89.   **HEDWIGIA,** *Ehr.*

420. **H. ciliata,** *Dicks.* Monoicous. Dichotomously branched, rooting at base only. L. crowded, spreading, sometimes secund, ovate-lanceolate, concave, margin recurved below, apex diaphanous, prolonged to a blunt point and strongly toothed on each side; per. l. with

apex laciniate. Caps. immersed, globose, lid convex,
with a short beak. Calyptra conical, sometimes hairy.

Rocks in mountainous districts.   March.

North Wales, Arthur's Seat, &c.

Var. β. LEUCOPHÆA. L. more crowded and spreading,
wider, and with longer diaphanous points.

Var. δ. VIRIDIS. L. scarcely secund, spreading, deep
green, scarcely diaphanous at apex.

Var. ε. STRIATA. L. plicate, much recurved, lid
conical.

90. HEDWIGIDIUM, B. and S.

421. **H. imberbe,** *Smith.* St. 1—3 inches, irregularly
not dichotomously branched, flagelliferous. L. ovate-
lanceolate, acuminate, imbricate when dry, margin
recurved, apex not diaphanous, but slightly crenate.
Caps. exserted, on a short seta, spherical or obovate,
lid with a blunt, slightly oblique beak. Calyptra
cucullate, reddish.

Rocks.   October, November.

Wales and Ireland.

Fam. 2. Cryphaeæ.

91. CRYPHÆA, *Mohr.*

422. **C. heteromalla,** *Hedw.* St. 1 inch, decumbent,
sparingly branched, subpinnate. L. spreading, imbricate, slightly recurved, broadly ovate, pointed, concave,
thickly nerved nearly to apex; per. l. elliptic, with an
excurrent nerve. Caps. oblong, immersed, appearing
secund, lid conical, pointed. Synoicous.

Trunks of trees.

Var. β. AQUATILIS. St. elongate. L. roundish, ovate,
obtuse.

Stones in running streams, or trees.   May, June.

Section III. *PLEUROCARPI.*

Tribe xxv.   LEUCODONTACEÆ.

Fam. 1. · **Leptodontæ.**

92.   LEPTODON, *Mohr.*

423. **L. Smithii,** *Dicks.*   St. 1—3 inches, creeping. Branches pinnate or bipinnate.   L. ovate, rounded and obtuse at apex, bisulcate, entire, margin recurved below on one side, nerved more than half way, apical cells small, rotund, basal, rhomboid, rectangular; per. l. erect, ovate, subulate. Caps. elliptical, horizontal, on a short, curved seta, lid with a straight beak.

Trees.   April.

Devon, &c.

Fam. 2.   **Leucodontæ.**

93.   PTEROGONIUM, *Swartz.*

424. **P. gracile,** *Dill.*   Rhizome creeping, with arcuate stems, and incurved, fasciculate branches. L. spreading (appressed when dry), ovate, concave, acute, serrate near apex, margin not recurved, slightly two-nerved at base. Caps. oblong, elongate, small-mouthed. Perist. fragile, lid conical, short, not rostrate.

Shady subalpine rocks and walls, and trees.   Spring.

94.   LEUCODON, *Schw.*

425. **L. sciuroides,** *Linn.* · St. or branches 1 inch, from a creeping rhizome. L. patent, when dry secund or imbricate, ovate-lanceolate, entire, tapering to a point, plicate, somewhat secund; per. l. longer, all nerveless. Caps. ovate-oblong or elliptical, erect, lid conical, beaked.

Trees, walls, rocks, &c.   Spring.

Var. β. MORENSIS. More robust. L. larger, densely conferted; per. l. larger, sheathing, pale. Caps. elongate-cylindric, subarcuate.

Craig Challeach (McKinlay), Berkshire, Surrey.

95. ANTITRICHIA, *Brid.*

426. **A. curtipendula,** *Linn.* St. 3—8 in., straggling, pinnately branched. L. ovate-lanceolate, concave, sharply curved to a roughly toothed point, which ends iu a double hook in the younger ones, nerved half way, margins recurved. Caps. roundish, elliptical, drooping, lid with a short, oblique beak.

Rocks and trees. April.

Tribe xxvi. NECKERACEÆ.

96. NECKERA, *Hedw.*

a. Capsule immersed.

427. **N. pennata,** *Hall.* St. 2 inches, pinnate, with complanate, longer branches. L. complanate, undulate, ovate-lanceolate, subcultriform, tapering to a slightly serrulate point, otherwise entire, nerveless, or sometimes shortly and faintly two-nerved. Caps. oblong or oval, immersed, lid with a short, oblique beak. Monoicous.

Trunks of trees. Rare. Spring.

b. Capsule exserted.

428. **N. pumila,** *Hedw.* St. 1—2 inches, subpinnate, with slender flagellæ and short, complanate branches. L. complanate, undulate, ovate-oblong, tapering, apiculate or acuminate, somewhat concave, serrulate, margin recurved, shortly two-nerved or nerveless. Caps. elliptical, erect, on a very short seta, lid with a short beak. Dioicous.

Trunks of trees and rocks. April, May.

429. [**N. Philippeana,** *Schp.* Primary stem creeping, densely pinnate, secondary ascending, remotely pinnate. L. densely imbricate, complanate, strongly and elegantly undulate, ovate-lanceolate, sharply narrowed into a longer or shorter flexuose apiculus, nerveless, areolæ small, linear. Bry. Eur., v. 445.] Probably only a barren form of *N. pumila.* Schp. Syn., ed. 2, p. 568.

Bark of a young ash tree.

Valley of Hirnant, Bala, N. Wales (Rev. H. H. Higgins), July, 1872, barren ; Scotland.

430. **N. crispa,** *Linn.* St. 4—6 inches, pinnate, from a creeping rhizome. L. complanate, undulate, ovate-oblong or ovate-lingulate, somewhat obtuse and pointed, serrulate at apex, faintly and shortly 2-nerved or singly nerved half way. Caps. ovate elliptical, erect, lid with a long, oblique beak. Dioicous.

Mountainous rocks, trees. Spring.

431. **N. complanata,** *Linn.* St. 1—2 inches, pinnate. Br. short, crowded, attenuate. L. complanate, not undulate, obliquely ovate-oblong, suddenly apiculate from broadish apex, faintly and shortly 2-nerved, apex serrulate. Caps. roundish, elliptical, tapering below, erect, lid large, obliquely rostrate. Dioicous.

Trunks of trees, walls, &c. Spring.

97. HOMALIA, *Brid.*

432. **H. trichomanoides,** *Schreb.* St. about 1 inch, irregularly pinnate. L. crowded, subsecund, subfalcate, complanate, oval, serrulate, apex obtuse, apiculate, faintly nerved half way. Caps. small, subcylindrical, suberect, lid with an oblique beak.

Trunks of trees and shady rocks.   October, Nov.

100. PTERYGOPHYLLUM, *Brid.*

435. **P. lucens,** *Linn.*   St. 1—3 inches, procumbent, with irregular complanate branches. L. complanate, large, roundish ovate, obtuse, entire, nerveless, areolæ large, hexagonal, pellucid. Caps. roundish, elliptical, almost pendulous, lid conical, suddenly tapering into a long, straight beak. Monoicous.

November, December.

Moist banks, stones in streams, &c.

Trunks of trees and rocks. April, May.

429. [**N. Philippeana,** *Schp.* Primary stem creeping,

Page 187. No. 433.

Thamnium alopecurum, *Linn.* Var. β. ANGUSTIFOLIUM, *Holt MS.* L. narrow, ligulate, margin more coarsely serrate, nerve broader, flatter, occupying one-third width of leaf.

By calcareous spring (with the type).

Ravensdale, Derbyshire, May, 1883 (G. A. Holt).

erect, lid large, obliquely rostrate. Dioicous.

Trunks of trees, walls, &c. Spring.

97. **HOMALIA,** *Brid.*

432. **H. trichomanoides,** *Schreb.* St, about 1 inch, irregularly pinnate. L. crowded, subsecund, subfalcate, complanate, oval, serrulate, apex obtuse, apiculate, faintly nerved half way. Caps. small, subcylindrical, suberect, lid with an oblique beak.

Trunks of trees and shady rocks.   October, Nov.

98. THAMNIUM, *Schp.*

433. **T. alopecurum,** *Linn.* St. 2—3 inches, naked below, pinnately branched above. L. spreading, ovate-lanceolate, acuminate, somewhat concave, serrate, strongly nerved nearly to apex. Caps. shortly ovate, cernuous or erect, lid with a long, oblique beak. Dioicous.

Moist woods, rocks, &c.   November.

Tribe xxvii. HOOKERIACEÆ.

Fam. 1. **Daltoniæ.**

99. DALTONIA, *Hooker and Taylor.*

434. **D. splachnoides,** *Smith.* St. ¼ inch, tufted, erect. Br. fastigiate. L. crowded, suberect, linear-lanceolate, acuminate, slightly keeled, entire, nerve vanishing below apex ; per. l. small, ovate. Caps. small, oval, oblong, suberect, lid large, with a long, straight beak.

October, November.

Subalpine moist shady rocks and trees.   Rare.

Fam. 2. **Hookeriæ.**

100. PTERYGOPHYLLUM, *Brid.*

435. **P. lucens,** *Linn.* St. 1—3 inches, procumbent, with irregular complanate branches. L. complanate, large, roundish ovate, obtuse, entire, nerveless, areolæ large, hexagonal, pellucid. Caps. roundish, elliptical, almost pendulous, lid conical, suddenly tapering into a long, straight beak. Monoicous.

November, December.

Moist banks, stones in streams, &c.

101. HOOKERIA, *Tayl.*

436. **H. lætevirens,** *H. and T.* St. shorter and more
slender, procumbent, subpinnate. L. complanate,
loosely imbricate, smaller, ovate, suddenly.and shortly
acuminate, with a thickened border, doubly nerved
above half way, serrulate at apex, areolæ smaller,
hexagonal. Caps. smaller, drooping, roundish, elliptical, lid as in 435. Monoicous. November, December.
Caves, wet rocks, and by rivulets.

Tribe xxviii. FABRONIACEÆ.

102. HABRODON, *Schp.*

437. **H. Notarisii,** *Schpr.* (*Pterogonium perpusillum,*
De Not.). St. creeping, irregularly branched. L.
spreading, squarrose, opaque, imbricate and shining
when dry, from an ovate base, longly acuminate, nerveless, entire, cells in narrow part elliptic fusiform, in
broader part quadrate, transversely rectangular; per. l.
internal with erose margins. Caps. oval-oblong, erect,
slightly striate and contracted at mouth when dry, lid
conical, erose. [Supp. Bry. Eur., fasc. iii. iv.] Syn.
ii. 588.
Trunks of elm and white thorn. Spring.
Windermere and Devon (J. Nowell); Killin, Perthshire (A. McKinlay), July, 1865; Ben Lawers (Hunt).

103. MYRINIA, *Schp.*

438. **M. pulvinata,** *Wahl.* (*Leskea,* ed. 1). St. ½ inch,
procumbent, slenderly branched. L. imbricate, elliptic-
ovate, inequilateral, narrowed below, concave, entire,
nerved half way, or 2-nerved at base, areolæ large,
rhomboid, quadrate at base; per. l. nerveless. Caps.

almost erect, oval, oblong, lid conical, pointed, falling early. Monoicous.

Roots of trees near rivers, &c. May, June.

Tribe xxix. LESKEACEÆ.

Fam. 1. **Leskeæ.**

104. MYURELLA, *Schp.*

439. **M. julacea**, *Villars* (*L. moniliformis*, Wahl., ed. 1). St. ½ inch, slender tufted, branched. L. imbricate, nearly round, obtuse, rarely shortly apiculate, very concave, nerveless, denticulate at base; per. l. ovate-lanceolate. Caps. almost erect, small, oval-oblong, with a short beak.

Alpine rocks. Summer.

England, Scotland, Ireland.

440. **M. apiculata**, *Hueb.* Loosely cæspitose, soft and fragile. L. loosely imbricate or patent, ovate, very concave, often with a recurved apiculus, opaque. Perist. pale, small.

Moist rocky ground. Summer.

Ben Lawers.

105. LESKEA, *Hedw.*

441. **L. polycarpa**, *Ehr.* St. intricately and densely tufted, almost pinnately branched. L. imbricate, subsecund, ovate-lanceolate, concave, tapering, stoutly nerved nearly to apex, margin entire, reflexed below; per. l. erect, striate. Caps. cylindrical, erect, lid conical, acute. Monoicous.

Roots of trees. May, June.

Var. *β.* PALUDOSA. Generally more lax. Caps. longer.

442. **L. nervosa**, *Schw.* Densely cæspitose. Br. pinnate. L. densely conferted, patent, imbricate when

dry, often subsecund, narrowly lanceolate from an
ovate base, plano-concave, margin reflexed, nerve
strong, cells close; per. l. int. sheathing, long apiculate, subsulcate.   Caps. narrowly oblong, and elongate-
cylindric, regular, solid, lid conical or shortly rostrate,
annulus narrow.   Perist. small, ext. teeth linear-
lanceolate, whitish, int. processes subulate, irregular,
pale, from a narrow basal membrane, shorter than the
teeth.   Dioicous.

Trunks of trees in alpine places.   Summer.
Scotland.

106.   ANOMODON, *H. and T.*

443. A. longifolius, *Schleich.*   Rhizomes slender, sec.
st. erect.   Br. subfasciculate.   L. somewhat secund,
from an ovate base, lanceolate, tapering, very acute,
bisulcate at base, nerved to apex.   Caps. oblong,
cylindric, on a short seta, annulus absent, lid large,
conical, rostellate.   Autumn.

Scotch mountains.   Yorkshire.   Fruit not found in
Britain.

444. A. attenuatus, *Schreb. (Hypnum, Schreb., Leskea,*
Hedwig.).   St. 1—2 inches, procumbent, with incurved
branches.   L. imbricate, sometimes secund, ovate-
lanceolate from a broader base, acute, scarcely nerved
to apex, areolæ minute, opaque; per. l. narrower,
nerveless.   Caps. cylindrical, erect, exannulate, lid
conical, with an oblique beak.   Dioicous.

Damp rocks and rotten tree trunks.

Den of Airlie, Forfarshire (Fergusson), 1868.

445. A. viticulosus, *Linn.*   St. 1—2 inches, from a
creeping rhizome, rigid.   L. more or less secund and
falcate-secund, from an ovate base, lingulate or subu-

late, obtuse, entire, nerve pellucid, almost reaching apex. Caps. almost cylindrical, erect, on a yellowish seta, lid narrowly conical, rostrate.

Shady limestone rocks, trees, &c. Spring.

Fam. 2. Pseudoleskeæ.

107. PSEUDOLESKEA, *Br. and Sch.*

446. P. atro-virens, *Dicks* (*Hypnum*, ed. 1). St. prostrate, irregularly branched, the latter slightly incurved, rigid, suberect. L. imbricate, somewhat secund, ovate-lanceolate, concave, with long, tapering points, margin recurved, entire or subserrulate above, thickly nerved almost to apex. Caps. oval or oblong, curved and subcernuous, lid convexo-conical.

Alpine rocks, &c. Rare in fruit. Spring. Scotland.

447. P. catenulata, *Brid.* St. about ½ inch, creeping, with erect, very slender branches. L. very small, ovate, acute, entire, concave, margin recurved, broadly nerved half way or more, basal marginal cells oblique. Caps. oval-oblong, slightly curved, suberect or cernuous, lid large, with a distinct beak. Dioicous.

Alpine and subalpine rocks. Summer. Scotland, Yorkshire, &c. Fr. not known in Britain.

Fam. 3. Thuidieæ.

108. HETEROCLADIUM, *B. and S.*

448. H. dimorphum, *Brid.* St. 1—2 inches, procumbent. Br. very slender, more or less pinnate; st. l. obcordate-acuminate, spreading, recurved; br. l. ovate, concave, obtuse, suberect, all serrulate and shortly 2-nerved, areolæ quadrate on margin, the rest larger, longer and less opaque. Paraphyllia leaf-like or pal-

mate. Caps. oblong, almost horizontal, lid conical, without beak. Dioicous.

Ben Lawers. Barren.   Autumn.

449. **H. heteropterum,** *Bruch.* St. procumbent, more or less pinnate, often rooting at apex; st. l. from an obovate base, acuminate ; br. l. ovate-acuminate, small, all more or less secund, serrulate, somewhat papillose at back, nerved singly half way, or short and forked. Paraphyllia ovate-acuminate, serrate. Caps. oblong, scarcely curved, almost horizontal, lid with a long beak. Dioicous.

Moist rocks near waterfalls.   November.
Ireland, Wales, Lancashire, &c.

109. **THUIDIUM,** *Schp.*
a. *Gracilia.*

St. prostrate. Br. pinnate and bipinnate. St. l. cells all rotund-hexagonal. Caps. horizontal, lid con- vexo-conical. Monoicous.

450. **T. decipiens,** *De Not.* (*H. rigidulum,* Ferg.). St. 2—4 inches, rigid, villous, with short, attenuate branches; st. l. distant, subsquarrose, deltoid-ovate, suddenly acuminate, auricled, concave, serrate, papillose on both sides, sometimes secund above ; br. l. smaller, crowded, ovate or ovate-lanceolate, spreading or secund, nerved half way or more, areolæ large, hexagonal, and pellucid at base, above oval-elongate, confused (Fer- gusson).

Springs and streams.

Ben Lawers (Dr. Stirton), 1866; Clova, 2800 feet (Fergusson), 1868; Auchinblae, 800 feet (Sim and Fergusson) ; Glas Mheal, Perthshire, 2500 feet (G. E. Hunt). All barren.

b. *Tamariscina.*

St. long, creeping, divided. Branches bi-tripinnate. Caps. elongate-cylindric, lid shortly rostrate.

451. **T. tamariscinum,** *Hedw.* St. elongate, arched, procumbent, interruptedly tripinnate. Br. not rooting; st. l. deltoid-obcordate, acuminate, plicate ; br. l. ovate, obtuse, all papillose at back, margin recurved, crenulate, irregularly serrulate near, and nerved almost to, apex. Caps. oblong-cylindrical, curved, cernuous, purplish-red, lid large, conical, with a long, thick beak.

Woods and banks. Frequent. November.

452. **T. recognitum,** *Hedw.* (*delicatulum,* L., ed. 1). St. elongate, erect or procumbent, bipinnate. Br. drooping, often rooting at apex; st. l. triangular; br. l. broadly cordate or ovate acuminate, concave, substriate, densely papillose or muricate on back and keel, finely serrulate, nerved nearly to apex, which is crowned with acute papillæ. Caps. subcylindrical, curved, cernuous, pale brown, lid large, conical, not rostrate.

Limestone and chalk rocks, &c. July, August.

c. *Abietina.*

St. erect. Br. simply pinnate, lid convexo-conical.

453. **T. abietinum,** *Linn.* St. 2—4 inches, rigid, reddish, not always erect. Br. slightly drooping, crowded. L. imbricate, erecto-patent, more or less secund; st. l. ovate or cordate acuminate, serrulate near apex, deeply sulcate, one margin plane, the other reflexed; br. l. narrower, less plicate, margin irregularly denticulate, all papillose on back and keel, nerved nearly to apex, areolæ dot-like. Paraphyllia filiform, mixed with tomentum. Capsule oblong-cylindrical, arcuate, cernuous, lid conical. Dioicous.

Alpine rocks, chalk hills, &c. May, June.

454. **T. Blandovii,** *W. and M.* St. 3 inches, erect, flexible. Br. slender, spreading. L. loosely imbricate, erect, from a spreading base, broadly ovate or sub-cordate, acute, narrowed at base almost to a pedicel, keeled, serrulate, not papillose on keel, thinly nerved nearly to apex, margin recurved, areolæ larger, sub-hexagonal. Paraphyllia laciniate-ciliate, mixed with tomentum. Caps. oblong, curved, cernuous, lid conical, with a blunt point. Monoicous.

Bogs. Rare. May.

Tribe xxx. HYPNACEÆ.

Fam. 1. Pterygynandreæ.

110. PTERYGYNANDRIUM, *Hedw.*

455. **P. filiforme,** *Timm.* St. creeping, with incurved, fasciculate branches. L. imbricate or secund, obovate and oblong, shortly acuminate, concave, papillose at back, serrulate at pointed apex, margin recurved, nerved half way, or shortly 2-nerved at base. Caps. elliptical, erect, lid conical, obliquely rostrate.

Mountainous rocks and tree trunks. May, June. Scotland, Ireland.

Fam. 2. Orthotheciæ.

111. LESCURÆA, *Schp.*

456. **L. saxicola,** *Milde.* Pr. st. creeping, regularly pinnatedly branched at extremity. L. subsecund, oblong-oval, shortly acuminate, apex distinctly serrate, concave, sulcate on one side, margin recurved, nerve vanishing below apex. Paraphyllia filiform. Caps. oval, oblong, on a long seta, lid conical. Perist. teeth entire or bifid, processes punctulate.

Stones in subalpine regions.   June.
Ben Lawers, August, 1880 (W. West).

112. CYLINDROTHECIUM, *Schp.*

457. **C. concinnum**, *De Not.* (*Montagnei*, Bry. Eur., and ed. 1). St. 1—2 inches. Branches pinnate, recurved, cuspidate. L. broadly ovate and ovate-oblong, concave, entire, rather obtuse, faintly 2-nerved at base, margin recurved below, incurved above, marginal basal cells large and pellucid. Caps. cylindrical, erect, on a long seta, and with a conical, blunt lid. Dioicous.

Limestone hills.   Autumn.
Scotland, England.

113. CLIMACIUM, *W. and M.*

458. **C. dendroides**, *W. and M.* St. 1—3 inches, erect, with long, spreading branches. L. ovate-lanceolate, serrulate at apex; st. l. acute; br. l. obtuse, nerved nearly to apex; per. l. nerveless, entire. Caps. erect, ovate-oblong, with a pointed beak. Perist. teeth united in a cone, incurved when dry.

Boggy places.   October—January.

114. ISOTHECIUM, *Brid.*

459. **I. myurum**, *Pollich.* St. 1—2 in., from a creeping, stoloniferous rhizome, with fasciculate branches. L. ovate-oblong or elliptical, concave, shortly acuminate, serrulate at apex, nerved half way, singly or forked; per. l. erect. Caps. ovate, erect, with a long, rostrate lid.

Trees and rocks.   Spring.

115. ORTHOTHECIUM, *Schp.*

460. **O. intricatum**, *Hartm.* (*Leskea subrufa*, Wils., ed. 1). St. about 1 inch or less, tufts intricate, fascicu-

lace, branched; foliage, young green, older brownish.
L. subsecund, lanceolate, long tapering, coucave, obsoletely sulcate, entire. Caps. ovate, tapering below,
almost erect, lid conical, pointed. Perist. int. processes
longer than teeth. Dioicous. Fr. not found in Britain.
  Subalpine rocks.   Summer.

  461. O. rufescens, *Dicks.* St. 1—3 inches, erect,
branched, with reddish-brown foliage. L. imbricate,
almost erect or subsecund, lanceolate, long tapering,
often almost piliferous, distinctly sulcate, margins
plane. Caps. generally erect, cylindrical, on a long,
smooth seta, lid shortly conical, int. processes with
cilia. Dioicous.
  Moist shady alpine rocks.   Summer.
  Scotland.

116. PYLAISIA, *Schp.*

  462. P. polyantha, *Schreb.* St. short, creeping,
branched. Branches incurved above. L. crowded,
subsecund, ovate-acuminate, suddenly apiculate, with
apiculus sometimes slightly serrulate, nerveless. Caps.
elliptic-oblong, erect, broader below, with a conical,
bluntish lid, per. processes cleft in the carina, rarely
bifid. Monoicous.
  Trees.   Autumn, winter.
  England, Scotland, Wales.

117. HOMALOTHECIUM, *Schp.*

  463. H. sericeum, *Linn.* St. 1 inch or more, creeping, branched. Branches erect, curved. L. imbricate,
• subsecund, lanceolate, long tapering, scarcely nerved
to apex, finely serrulate, areolæ very narrow. Caps.
almost erect, cylindrical, tapering above, on a rough
seta, lid conical, obliquely beaked. Perist. teeth with
a hyaline margin. Dioicous.   November—March.

Walls, rocks, and trunks of trees.

Fam. 3. **Camptotheciæ.**

118. **CAMPTOTHECIUM**, *Schp.*

464. **C. lutescens**, *Huds.* St. about 3 inches, irregularly branched, sometimes pinnate. L. narrowly elongate-lanceolate, tapering to a long point, entire, minutely serrulate at apex, nerved nearly to apex. Caps. oblong, arcuate, on a rough seta, lid conical, beaked. Monoicous.

Rocks and woods (limestone and sandstone). April.

465. **C. nitens**, *Schreb.* St. 2—4 inches, erect, almost pinnate, radiculose. Branches short, spreading. L. erecto-patent, lanceolate, subulate, acuminate, not nerved to apex, entire, margin recurved. Caps. arcuate, oblong, on a long, smooth seta, lid conical, ext. per. yellow. Dioicous.

Bogs.   April—June.
England, Scotland.

Fam. 4. **Brachytheciæ.**

119. **PTYCHODIUM**, *Schp.*

466. **P. plicatum**, *Schl.* St. creeping, tomentous· Branches ascending, incurved. L. erecto-patent or subsecund, ovate-lanceolate, long pointed, margin recurved, apex sometimes obsoletely denticulate ; per. l. twice as large. Caps. ovate-oblong, small, on a smooth, reddish seta, lid convexo-conical, apiculate. Dioicous.

Alpine rocks.   Scotland.   Winter.

120. **BRACHYTHECIUM**, *Schp.*

a. Seta smooth.

467. **B. salebrosum**, *Hoffm.* St. radiculose, 1—2 in.,

procumbent, subpinnate. L. ovate-lanceolate, acute or
filiform acuminate, serrulate, more or less deeply
striate, nerved more than half way. Caps. ovate,
cernuous, curved, lid conical, scarcely beaked, or conical
apiculate. Monoicous.

Trees.   Autumn.

Near Kirkham Abbey, Yorkshire (R. Spruce);
Sussex (Mitten); Scotland, &c.

Var. γ. MILDEANUM, *Schp.* (*palustre*, Sch. Syn., ed. 2,
p. 641). St. upright, tall, robust, not pinnate, and
without radicles. L. broader, more concave, less distinctly sulcate, apiculus shorter.

Damp places.

Sands at Southport, Cornwall, Dublin, Fifeshire, &c.

468. **B. glareosum,** *Bruch.* St. about 2 inches, sub-
procumbent. Branches sometimes subpinnate. L.
erecto-patent, from an ovate base, gradually tapering
into a long, slender, sometimes twisted, subserrulate
apex, margin reflexed below, interruptedly sulcate,
thinly nerved more than half way. Caps. ovate-oblong,
cernuous, arcuate, lid conical, with a distinct beak.
Dioicous.

Woods and shady banks.   November.

469. **B. albicans,** *Neck.* St. about 2 inches, upright.
L. spreading, appressed when dry, ovate and oblong-
lanceolate, subpiliferous, sulcate, concave, entire, or
obsoletely serrulate at apex, nerved more than half
way. Caps. ovate, small, scarcely curved, cernuous, on
a slender seta. Young foliage pale green, greyish-
brown below. Dioicous.

Sandy, grassy places.   Spring.

b. Seta rough.

470. **B. velutinum,** *Linn.* St. short, creeping, with

erect, irregularly pinnate branches. L. subfalcate, secund, ovate-lanceolate, prolonged into a taper point, nerved half way or more, margin serrulate, reflexed below; per. l. almost piliferous. Caps. roundish ovate, cernuous, lid convexo-conical, muticous. Monoicous.

November, March.

Walls, sandy hedge banks, roots of trees, &c.

471. **B. reflexum**, *W. and M.* St. more or less arched, procumbent, and rooting at extremities. Branches subpinnate, slender, incurved. L. from a broad, ovate-cordate base, more or less suddenly acuminate, almost piliferous, finely serrate, nerved almost or quite to apex, margin reflexed, areolæ lax, quadrate, excavate at basal angles. Caps. small, roundish-ovate, horizontal, lid conical, pointed. Monoicous.

Scottish mountains. Autumn.

472. **B. Starkii**, *Brid.* St. creeping or densely cæspitose, ascending. Branches erect, arcuate at summit. St. l. obcordate-lanceolate; br. l. ovate-lanceolate, gradually acuminate, acumen slightly twisted, serrate, nerved half way, cells hexagono-rhomboid; per. l. squarrose, nerveless. Caps. turgid-ovate, subglobose, horizontal, lid convexo-conical, annulus broad. Per. processes with cilia. Monoicous.

Stones and ground in woods. September—March. Scotland, Ben Lawers, &c.

473. **B. glaciale**, *B. and S.* St. creeping, branched. L. imbricate, julaceous, ovate-lanceolate, more or less apiculate, slightly sulcate, serrate, nerved nearly to apex; per. l. imbricate, nerveless, with a long, flexuose apiculus. Caps. cernuous, horizontal, oval or ovate, lid convexo-conical, with a short beak, annulus narrow. Monoicous.

Alpine rocks, stones, &c.   August, September.
Ben Challum (McKinlay).

474. **B. rutabulum**, *Linn.* St. long, loosely tufted, procumbent, and rooting at extremities, with erect branches. L. patent, broadly ovate, lanceolate, acuminate, serrulate, striate when dry, thinly nerved above half way, basal cells broad, shortly hexagonal at base. Caps. ovate oblong, arcuate, cernuous, on a very rough seta, lid bluntly pointed. Monoicous.

September—March.
Banks, walls, and trees. Common.

475. **B. campestre**, *B. and S.* St. loosely cæspitose, prostrate or ascending, much branched. L. erecto-patent, longly ovate-lanceolate, more or less subulato-acuminate, serrulate, thinly nerved more than half way, plicate, shining; per. l. recurved, squarrose from the middle, piliferous. Caps. oblong-cylindrical, subarcuate, on a slightly roughened seta.

Grassy places, fields, &c.   Winter and spring.
Maresfield, Sussex (Mr. Mitten); Spec. in Herb. Kew. "Newchurch, Over, Cheshire, W. W., Dec. 13, 1837."

478. **B. rivulare**, *Bruch.* St. arched, slender. Br. slender, incurved, subpinnate. L. deltoid-ovate, gradually tapering, not suddenly acuminate, serrate, nerved above half way, decurrent, margin plane or recurved. Caps. short, roundish ovate, arcuate, cernuous, lid large, conical, acute, rostellate. Dioicous.   Autumn.
Stones, &c., by rivulets in shady woods; sometimes in water, when the stems are often very elongate.

479. **B. populeum**, *Hedw.* St. creeping, subpinnate. L. narrowly ovate-lanceolate, tapering to a long, serrulate point, concave, margin plane, nerved to apex.

Caps. small, roundish ovate, slightly cernuous or nearly erect, seta smooth below, roughish above, lid conical, very acute, subpersistent on the ripe fruit. Monoicous.' September – February.

Walls, rocks, trees, &c. Frequent.

Var. β. MAJUS. More robust. L. closer, longer, straighter.

480. **B. plumosum,** *Swartz.* St. creeping. Branches long, frequently erect, subpinnate. L. broad, ovate, concave, shortly and obliquely acuminate, subsecund, entire or serrulate near apex, nerved above half way. Caps. small, roundish ovate, cernuous, seta roughish at summit only, lid conical, acute. Monoicous.

October—March.

Subalpine shady rocks, stones in rivulets, walls, &c.

Fruit unknown.

481. **B. cirrhosum,** *Schwg.* St. erect or procumbent, with a few erect branches. L. tumid, imbricate, elliptic, obovate-oblong, subcochleariform, entire, except near the long, narrow points, which are serrulate and suddenly geniculate or reflexed where the point joins the blade, concave, nerved halfway. Never been found in fr.

Summit of Ben Lawers, 1823 (Dr. Arnott, G. E. Hunt).

121. **MYURIUM,** *Schp.*

482. **M.** **Hebridarum,** *Schp.* (*Leucodon Lagurus,* var. *borealis,* Wils. Bry. Brit.). St. not tomentous. Br. irregular, erect. L. auriculate and serrate at base, obovate-oblong, margin incurved, cucullate, concave, and suddenly attenuated into a long, lanceolate-filiform apiculus, nerveless, remotely serrate below, above densely serrulate.

N. Uist, Hebrides, 1851.   More recently by Mr. J. Shaw.

122. SCLEROPODIUM, *Schp.*

483. **S. cæspitosum,** *Wils.*   St. densely tufted, creeping.   Br. slender, short, incurved.   L. subsecund, ovate-lanceolate, acuminate, small, concave, serrulate, margin recurved, nerved above half way, nerve rarely bifurcate.   Caps. subcylindrical, slightly arcuate, suberect, lid long, rostellate.

Damp walls and roots of trees.   Autumn—spring.

Lancashire, Yorkshire, Cheshire, Sussex, Scotland, &c.

484. **S. illecebrum,** *Schwaeg.*   St. procumbent, sometimes subpinnate.   Br. incurved, obtuse.   L. erecto-patent, roundish-ovate, pointed, very concave, imbricate, serrulate, tip slightly recurved, nerve reaching above half way, its tip slightly projecting from back of leaf, rarely bifid.   Caps. ovate-oblong, cernuous, somewhat ventricose, lid bluntly conical, apiculate.

Banks and rocks near the sea.   November, December.

Hampshire, Anglesey, &c.

123. HYOCOMIUM, *Schp.*

485. **H. flagellare,** *Dicks.*   St. 1 inch or more, arched, pinnate.   Br. subfasciculate, recurved ; st. l. squarrose, broadly cordate-acuminate, apiculate, apiculus recurved, slightly striate ; br. l. less spreading, subsecund, roundish-ovate, less acuminate, all sharply serrate and mostly 2-nerved at base or nerveless; per. l. almost erect, much narrower, nerveless.   Caps. ovate-oblong, curved, cernuous, on a rough seta, lid convex, apiculate.   Dioicous.   October, November.

Moist shady rocks by cascades, &c.

124. EURHYNCHIUM, *Schp.*

Section I. *STRIATA.*

Leaves more or less distinctly sulcate or striate.

a. *Seta smooth.*

486. **E. myosuroides**, *L.* (*Isothecium myosuroides*, Bry. Brit., 323). St. slender, creeping, secondary stem erect, dendroid. Br. fasciculate, incurved. L. somewhat spreading, from an ovate or cordate base, lanceolate acuminate, serrulate, nerved more than half way; br. l. narrower, lanceolate. Caps. elliptic-oblong, more or less inclined, slightly curved, on a twisted or curved seta, lid conical, with a short beak. Dioicous.

Trunks of trees and rocks. Spring.

487. **E. strigosum**, *Hoffm.* St. 1 inch, suberect or creeping, scarcely pinnate. L. roundish-ovate or cordate, concave, rather obtuse, serrate, nerved above half way, margin recurved below. Caps. subcylindrical, curved, small, lid conical, with a longish, curved beak. Parasitico-monoicous.

Roots of trees, rocks, &c. November.
Cornwall (Tozer in Herb. Hook.). Spring.

488. **E. circinnatum**, *Brid.* St. short, suberect, arched. Br. curved and drooping. L. very small, ovate-lanceolate, upper lanceolate, subsecund, serrulate, thinly nerved nearly to apex, areolæ oval, smaller and quadrate at base. Caps. oblong, cernuous, curved, lid large, with a long, oblique or curved beak. Dioicous.

Shady limestone rocks and walls. March.

489. **E. striatulum**, *Spruce.* St. short, creeping, tufted. Br. short, crowded, erect. L. erecto-patent, broadly cordate, lanceolate, long taper-pointed, serrate, substriate, strongly nerved more than half way, basal

areolæ opaque, minute. Caps. oblong, cernuous, lid roundish, with a long, pointed beak. Dioicous.

Shady limestone rocks and roots of trees.   Spring.

490. **E. striatum**, *Schreb.* Much larger than the last in all its parts. St. loosely tufted, arched, subpinnate. Br. drooping. L. gradually tapering from a broad, cordate base (br. l. obovate-lanceolate), almost squarrose, serrate, striate, nerved more than half way. Paraphyllia numerous, rotund-ovate. Caps. almost cylindrical, curved, cernuous, lid large, with a long, slender, curved beak. Dioicous.

Woods and shady banks.   December.

b. *Seta rough.*

491. **E. crassinervium**, *Tayl.* St. creeping. Br. erect. L. spreading, ovate-lanceolate, suddenly acuminate, apiculate, serrate, concave, margin reflexed, nerve thick, reaching nearly to apex, sometimes forked. Caps. elliptic-oblong, small, curved, cernuous, lid large, with a very long, slender, oblique beak. Dioicous.

Shady limestone or quartzose rocks.   Spring.

492. **E. piliferum**, *Vaill.* St. 2—3 inches, slender, procumbent, branched. L. imbricate, erecto-patent, elliptical, serrulate, suddenly contracted into a long, serrulate, almost piliferous point, concave, nerved half way; br. l. smaller. Caps. oblong, cernuous, slightly arcuate, lid with a long, curved beak.

Shady banks and woods. Fruit rare.   Spring.

Section II. *PRÆLONGÆ.*

L. opaque, scarcely sulcate, cells broader and shorter.

a. *Seta rough.*

Synoicous.

493. **E. speciosum**, *Brid.* Loosely cæspitose. Br.

irregular or subpinnate. L. spreading, ovate-lanceolate, almost flat, thinly nerved, serrulate, cells long, narrow; per. l. squarrose, lower nerveless, upper oblong, linear-lanceolate, nerved. Caps. turgidly ovate, curved, cernuous or horizontal, lid convex, long-beaked.

Stones and roots in wet places. Autumn.

Porth Dafarch, Holyhead (Wilson, 1830); Sussex (Mitten): near Penzance (Curnow); Ireland, &c.

494. [**E. hians**, *Hedw.* St. prostrate, extended, irregularly branched. Br. ascending. L. spreading, shining, ovate-cordate, serrulate, nerve uniform, ceasing about middle. Caps. cernuous, lid beaked as long as capsule.]

Woods, among decayed leaves.

Sussex (Mitten). Rev. J. Fergusson refers this to *E. Swartzii*, in litt.

Dioicous.

495. **E. Swartzii**, *Turn.* (*prælongum*, ed. 1, and Bry. Eur. et Schp. Syn.). St. long, arched or procumbent, often bipinnate, branches slender; st. l. squarrose, recurved, broadly cordate, and suddenly tapering to a long point, amplexicaul, nerve carried nearly to base of point; br. l. lanceolate, acuminate, all serrate; per. l. thinly nerved. Caps. small, oval-oblong, tumid, obliquely cernuous, lid with a long, tapering, slender beak.

Moist shady banks. November.

496. **E. pumilum**, *Wils.* St. creeping, filiform. Br. slender, subpinnate, subcomplanate. L. minute, ovate-lanceolate, spreading, subserrulate, faintly nerved half way; per. l. smaller, recurved. Caps. short, roundish, ovate, cernuous, on a short seta, lid rather large, with an oblique beak.

Shady rocks and hedge banks.   Spring.
England, Scotland, Ireland.

497. **E. Teesdalii,** *Sm.* St. slender, creeping. Br.
erect. L. subcomplanate, narrowly lanceolate, rigid,
slightly serrulate near apex, broadly nerved nearly to
apex. Caps. thickly obovate, cernuous, on a short
seta, lid almost as large as capsule, beaked.

Moist shady rocks near waterfalls.   March—June.

498. **E. prælongum,** *Dill.* (*Stokesii*, ed. 1, Bry. Eur.
et Schp. Syn.). St. densely cæspitose, branches ascending, simple below, above densely pinnate and bipinnate;
st. l distant, acutely cordate, shortly acuminate, and
triquetrous, recurved; br. l. ovate-lanceolate, erecto-
patent, all thinly nerved, and serrate. Paraphyllia
numerous, lanceolate, deltoid, erose. Caps. oblong,
ventricose, horizontal, olive-coloured, lid with a long,
straight, subulate beak, from a conical base.

Stones and rocks in woods, &c.   Autumn.

Var. *β.* STOKESII, *Turner.* More evidently bipinnate.
Br. crowded. L. cordate, acuminate, erecto-patulous.
Caps. ovate.

Rocks in shady mountainous places.

125. **RHYNCHOSTEGIUM,** *Schp.*

a. *Demissa.*

Low, slender. L. complanate, oblong-lanceolate,
entire, nerveless. Caps. thin walled, with a slenderly
beaked lid, seti smooth. Monoicous.

499. **R. demissum,** *Wils.* St. depressed, filiform.
Br. short, slender. L. elliptic-lanceolate, acute, subsecund above, margin entire, reflexed. Caps. small,
narrowly elliptical, horizontal, cernuous, lid obliquely
rostrate.

Shady mountainous rocks.   Winter.
North Wales, Ireland.

b. *Tenellæ.*

St. creeping. L. narrowly lanceolate. Caps. solid-
walled, seta rough or smooth. Monoicous.

500. **R. tenellum,** *Dicks.* St. and br. very short,
creeping. L. erecto-patent, narrowly lanceolate, acuminate, almost setaceous, light green, entire, nerved
nearly to apex. Caps. ovate, cernuous, on a smooth
seta, lid beaked.

Walls and rocks, principally limestone.   Spring.

501. **R. curvisetum,** *Brid.* Low and broadly cæspitose. Br. erect or spreading, irregular. L. erecto-
patent, remote below on stem and branches, clustered
at summits, and narrowly oblong-lanceolate, acuminate,
concave, nerved half way, margin more or less distinctly
serrate, basal cells hyaline. Caps. oval or oblong,
horizontal, seta roughish, flexuose.

Stones and rocks near streams.   Autumn.
South of England, Yorkshire.

c. *Depressæ.*

Plants low, cæspitose, soft. L. broadly oblong, thinly
nerved, seta smooth.

* Dioicous.

502. **R. depressum,** *Bruch.* St. prostrate, pinnate.
Br. thickest in middle—both very short. L. complanate, rarely subsecund, crowded, ovate-oblong, shortly
acuminate, slightly concave and finely serrulate, shortly
and faintly 2-nerved. Caps. oval-oblong, curved, cernuous, lid not as long as capsule, and long beaked.

Rocks and stones, especially limestone.   Autumn.

Caergwrle, Anglesey, October, 1871, in fruit (C. L. Higgins), &c.

** Monoicous.

503. **R. confertum,** *Dicks.* St. creeping, subpinnate. Br. erect. L. slightly secund or complanate, ovate, acuminate, concave, serrulate, thinly nerved quite or more than half way. Caps. ovate-oblong, cernuous, lid short, with a curved beak, as long as capsule.

Rocks, walls, trees, &c. Frequent. October, March.

504. **R. megapolitanum,** *Bland.* Much larger than the last, and remotely branched. St. l. remote ; br. l. crowded, lower oblong-lanceolate, the others more or less sharply acuminate from a broad, ovate base, slightly or sharply serrulate, nerved more than half way. Caps. oblong, cylindrical, incurved, arcuate when dry, lid with a short, thick beak.

Sandy shores.   Spring.
Southport, Dublin, Sussex, &c.

505. **R. murale,** *Hedw.* St. short, creeping, with erect, crowded branches. L. closely imbricate, round-ish, ovate, concave, faintly serrulate, cucullate at apex, which is slightly mucronate, not acuminate, nerved half way. Caps. ovate, somewhat cernuous, lid flattish, with a long beak.

Walls, &c., chiefly limestone.   Spring.

Var. β. COMPLANATUM. St. long, creeping, with fewer branches. L. complanate, scarcely concave.

Var. γ. JULACEUM. L. densely imbricate, obtuse, very concave, cochleariform, frequently rufescent.

506. **R. rusciforme,** *Weis.* St. creeping, with long, irregular, procumbent branches, sometimes floating. L. complanate and subsecund, ovate, with a cordate base or broadly oblong, acuminate, serrate, stoutly

nerved nearly to apex. Caps. shortly ovate, cernuous, lid convex, with a very long beak.   November.
Rocks and stones in rivulets. Frequent.

Fam. 5. **Hypnæ.**

126. PLAGIOTHECIUM, *Schp.*

a. Peristome teeth remotely articulate, cilia without
appendages (ciliolæ).

Dioicous.

507. **P. latebricola**, *Wils.* (*Leskea*, Bry. Brit.). St. short, slender, sparsely branched, suberect. L. laxly complanate or subsecund, ovate-lanceolate, tapering, acute, entire, slightly concave, decurrent, faintly 2-nerved or nerveless, margin recurved. Caps. ellipticoblong, turbinate when dry, lid acutely conical, large, apiculate.   Winter.
Moist shady woods, decaying trunks, and ferns.

b. Peristome teeth densely articulate, basal membrane
broad, processes with long cilia.

1. Leaves distichous, complanate, rarely complanatesecund.

508. **P. pulchellum**, *Hedw.* St. arcuate, short, densely tufted, suberect. Br. fastigiate. L. crowded, secund, subfalcate, lanceolate, gradually tapering from base to apex, entire, generally nerveless. Caps. oblong, suberect, curved, lid conical, muticous. Monoicous.
Mountainous shady rocks, &c.   June—October.
509. **P. nitidulum**, *Wahl.* Broadly low, cæspitose. St. prostrate. Br. erect. L. complanate, rarely subsecund, larger than last, broader and more shining, with a longer apiculus. Caps. cernuous, horizontal

when dry, ovate or oval, truncate when empty, lid convexo-conical, obtuse. Monoicous.

On decaying trunks and leaves.   Summer.

Heseltine Ghyll, Yorkshire, 1861 (Whitehead), Scotland. Rare.

510. **P. denticulatum**, *Linn.* St. prostrate, with subfasciculate branches. L. complanate; at apex and base of branches, small, broadly lanceolate; in middle larger, ovate oblong, apiculate, all inequilateral, sometimes subserrulate at apex, margin recurved below, shortly 2-nerved. Caps. arcuate, oblong, suberect or cernuous, lid acutely conical. Monoicous.

Subalpine woods, banks, wet rocks, &c.   Summer.

Var. β. OBTUSIFOLIUM.   L. elliptical, more or less obtuse, slightly concave.

Alpine rocks.

511. **P. Borrerianum**, *Spruce* (1846). St. prostrate. Br. distichous, fasciculate or pinnate. L. complanate, ovate-lanceolate, tapering to a long, slender, serrulate point, sometimes oblique, nerveless, or faintly 2-nerved. Caps. small, ovate, suberect or scarcely horizontal (not pendulous). Outer and inner peristome *pale yellow.*

(Our English moss is described in Bry. Brit. as *H. elegans*, Hook., which is an exotic species not occurring in Britain, having its leaves not tapering into a long, slender point, and capsule pendulous, with a *red*, outer peristome.)   April.

Shady banks and rocks, usually barren, but fruit has been gathered at Arthog, North Wales, first by Mr. Whitehead, and since at several other places sparingly.

Var. β. COLLINUM, *Wils.* St. erect, tufted. L. subsecund.

512. **P. sylvaticum**, *Linn.* St. longer, about 1 inch,

decumbent, branched. L. subcomplanate, sometimes subsecund, ovate-lanceolate, shortly acuminate, entire, obsoletely 2-nerved. Caps. cylindrical, curved, inclined or horizontal, sulcate when dry, lid long, shortly beaked. Dioicous.

Roots of trees in woods, &c. September.

Var. β. SUCCULENTUM. Todmorden.

513. **P. undulatum,** *Linn.* St. and br. procumbent, 2 inches or more. L. complanate, ovate, acute, not acuminate, entire, undulate, margin incurved on one side, faintly 2-nerved, whitish-green. Caps. cylindrical, tapering at base, cernuous or horizontal, striate when dry, lid with a short beak. Dioicous.

Woods and moist places. September.

2. Leaves spreading or spreading-secund.

514. **P. Mühlenbeckii,** *Schp.* St. short, tufted, sub-erect, with recurved, fasciculate branches. L. complanate, subsecund, spreading, deltoid-ovate or subcordate, tapering to a long, acuminate point, one side incurved, dark green, finely serrulate, nerveless or shortly 2-nerved. Caps. oblong, cylindric, slightly inclined, tapering below, striate when dry, lid short, conical. Monoicous. July.

Alpine rocks. Scotland, Ireland.

515. **P. Silesiacum,** *Seliger.* St. and br. procumbent, the latter arcuate. L. secund, mostly pointing upwards, ovate-lanceolate, long tapering, concave, distinctly serrulate, obsoletely and shortly 2-nerved. Caps. long, subcylindrical, not striate when dry, curved, cernuous, lid conical, pointed. Monoicous.

Stems of decaying trees, rocks, &c. Summer.

Kent, Yorkshire.

127. AMBLYSTEGIUM, *Schp.*

a. Leaves opaque, cells wholly parenchymatose, more or less chlorophyllose.

1. Dioicous.

516. **A. Sprucei,** *Bruch.* St. short, slender, with few branches. L. distant, narrowly ovate-lanceolate, long pointed, margin almost entire, concave, nerveless; per. l. larger, with longer points, distinctly serrulate at apex. Caps. erect, elliptical, turbinate when dry, mouth wide, lid mamillate. [*Leskea,* Bry. Brit.]

Shady subalpine rocks. Rare. Summer. Teesdale, Todmorden, &c.

2. Monoicous.

517. **A. confervoides,** *Brid.* St. creeping, very slender, subpinnate, sparingly branched. Br. capilliform. L. scattered, subsecund, more or less spreading, ovate-lanceolate, acuminate, entire, nerveless; per. l. longer, erect. Caps. cernuous, oval-oblong, slightly incurved, pale brown, semi-pellucid, lid convex, obliquely apiculate.

Stones in shady places, limestone. Summer. Dovedale (Dr. Fraser, 1866), Westmoreland, &c.

518. **A. serpens,** *Linn.** St. creeping, subpinnate, with slender, suberect branches; st. l. spreading, ovate-lanceolate; br. l. narrower, subsecund, tapering into long points, entire, faintly nerved half way, or sometimes nearly to apex. Caps. oblong, cylindrical or obovate, curved, cernuous, reddish at mouth, lid conical, acute. April, May.

* The Southport moss referred by Kindberg to *A. porphyrrhizon* belongs here.

Walls, moist banks, trees, &c. Common.

519. **A. radicale**, *P. Beauv.* St. creeping, with sub-erect, rigid branches. L. spreading, ovate-lanceolate, from a cordate or deltoid base, and strongly nerved almost to the long, tapering apex; per. l. larger, serrate, seta long (sometimes 2 inches). Caps. arcuate-cylindric, cernuous, not red at mouth, lid conical, with a short, sharp beak.

Moist ground amongst grass. April, May.

520. **A. irriguum**, *Wils.* St. procumbent, rigid, sometimes pinnate. L. spreading, secund, gradually tapering to a point, from a deltoid-ovate, somewhat decurrent base, subserrulate, strongly nerved nearly to apex. Caps. oblong, cernuous, curved, when dry more so, and contracted at mouth, annulus persistent, lid conical, blunt pointed.

Stones in rivulets and streams. April.

521. **A. fluviatile**, *Swartz.* St. procumbent, with simple, prostrate, not rigid branches. L. patent or subsecund, ovate or ovate-lanceolate, decurrent, acute, entire, concave, strongly nerved almost to apex, margin recurved below. Caps. slender, elliptical, elongate, suberect, only slightly curved, lid conical.

Rocks and stones in mountain streams. May, June.

b. Leaves broadly ovate, more or less longly acumi-nate, decurrent at angles, cells thin-walled, long hexagono-rhomboid, laxer at base, at angles rectan-gular.

522. **A. riparium**, *Linn.* St. long, creeping, with subpinnate, suberect branches. L. spreading, rarely subsecund, subcomplanate, ovate-lanceolate, entire, nerved two-thirds or more. Caps. oblong, cylindrical,

curved, cernuous or horizontal, contracted at mouth when dry, lid conical, pointed.   Monoicous.

May, June.

Stones, &c., near pools, sometimes in water.

128. HYPNUM, *Dill.*

Sub-genus 1.   *Campylophyllum.*

Pl. arched, creeping, strongly radiculose.   L. closely set, *divaricato-squarrose*, from a sheathing base, cells narrow, linear, quadrate at angles.

a. Monoicous.

523. **H. Halleri**, *L. fil.*   St. creeping, with pinnate, erect branches.   L. crowded, lanceolate, recurved from a roundish ovate, erect base, acuminate, serrulate, almost squarrose, faintly 2-nerved at base or nerveless. Caps. oblong, cylindric, curved, cernuous, lid conical, blunt, orange.

Alpine rocks.   Rare.   August.

Sub-genus 2.   *Campylium.*

St. creeping, prostrate or ascending.   Br. irregular or subpinnate.   L. *squarrose-divaricate* or subsecund, falcate, lanceolate, with a long acumen, nerveless or thinly nerved, cells narrow, linear, generally quadrate and lax at angles.   Caps. subarcuate, lid convexo-conical.

524. **H. Somerfelti**, *Myr.* (*polymorphum*, Hedw., ed. 1). St. procumbent, branches simple, erect, slender.   L. spreading, almost squarrose, subsecund, ovate, suddenly lanceolate, acuminate, with a long apiculus, entire, nerveless or with 2 striæ at base.   Caps. oblong, curved, cernuous, lid conical.   Monoicous.   June.

Limestone walls, banks and rocks, roots, &c.

525. **H. elodes,** *Spruce.* St. elongate, slender, with subpinnate, slender, suberect branches. L. distant, spreading; br. l. lanceolate-subulate, apex almost setaceous, secund; st. l. wider, less secund, all entire or subserrulate at base, nerved nearly or quite to apex. Capsule cylindrical, curved, cernuous, lid conical. Dioicous.

Wet places and bogs.   April, May.

526. **H. chrysophyllum,** *Brid.* St. creeping, pinnate. L. almost squarrose, subsecund, from a cordate, ovate, concave base, tapering into long setaceous points, entire, nerved more than half way, areolæ not enlarged or diaphanous at base. Caps. large, cylindrical, curved, cernuous, lid conical. Dioicous.

Fallow ground, chalk hills, &c.   May—September.

527. **H. stellatum,** *Schreb.* St. 1—2 inches, erect, densely tufted. Br. irregular or subpinnate, cuspidate. L. squarrose, recurved, rather suddenly tapering into a long point, from a deltoid-ovate base, with a few large, diaphanous cells at basal angles, nerveless, entire. Caps. oblong, curved, cernuous, lid convex, pointed. Dioicous.

Marshes and bogs.   May, June.

Var. β. PROTENSUM. St. creeping, much branched. L. shorter.

528. **H. polygamum,** *Bry. Eur.* St. 1 inch or more, procumbent, subpinnate. L. spreading, less squarrose, ovate-lanceolate, tapering into shorter points than last two, entire, nerved nearly to apex, areolæ larger at base. Caps. oblong, subcernuous or subhorizontal, lid conical, pointed.

Wet swampy places.   May.

Var. β. STAGNATUM. St. longer, suberect, more pinnate. L. with a longer nerve, seta longer, often 3 inches or more.

Sub-genus 3. *Harpidium.*

Br. generally pinnate. L. *falcate*, singly nerved, cells narrowly linear, laxer towards the base, enlarged and excavate at angles.

a. Dioicous.

† Stems and branches strongly hooked at apex.

529. **H. aduncum**, *Hedw.*, St. iv. t. 24 (*H. Kneiffii*, B. and S., ed. 1). St. 2—6 inches long, erect, subpinnate; st. l. falcato-secund, somewhat distant, oblong-lanceolate, with a long, flexuose acumen, occasionally faintly subserrulate near the base, thinly nerved two-thirds the length, not striate; br. l. smaller, subsecund or patent, basal angles decurrent, excavate, of lax, sub-quadrate cells, middle, hexagono-rectangular, apical, narrow, elongate-rectangular; per. l. erect, oblong-lanceolate, int. longly acuminate, and slightly sulcate. Caps. cylindrical, oblong, arcuate, broadly annulate.

Swamps and marshes.   June.

Var. β. KNEIFFII. St. long, flexuose, scarcely branched. L. patent or subsecund, suberect or somewhat falcato-lanceolate, ovate-sagittate, nerved to about middle, basal cells narrower; int. per. l. deeply sulcate.

530. **H. vernicosum**, *Lindb.*, 1861 (*H. pellucidum*, Wils. MS.; *H. aduncum*, var. *tenue*, Bry. Brit.). St. erect, rather rigid, pinnate. L. shorter, falcato-secund, the apical ones involute, ovate, oblong-lanceolate, more or less acuminate, distinctly sulcate, neither auricled nor decurrent, very glossy, yellow-green, nerve vanishing far below apex, cells very narrow, vermicular,

at base rotund, quadrate, and purplish. Caps. oblong, cernuous, arcuate, lid mamillate, annulus broad.

Wybunbury Bog, Cheshire, &c. June.

531. **H. Cossoni,** *Schp.* (*intermedium,* Lindb.). In habit like *H. Sendtneri,* var. β. St. elongate, flexuoso-erect, interruptedly pinnate. Br. very unequal. L. falcato-secund, ovate-oblong, becoming lanceolate, acumen subulate, scarcely furrowed, with minute, decurrent auricles, nerved nearly to apex, cells very narrow, vermicular, opaque, at basal angles rotund, quadrate, scarcely excavate; outer per. l. squarrose, inner almost erect, acuminate, sulcate. Fruit as in *Sendtneri.*

Bogs. Frequent.

532. **H. Sendtneri,** *Schpr.* (*H. aduncum,* ε, *hamatum,* and ζ, *giganteum,* Bry. Eur.). St. 3—6 inches, simple, pinnate. L. falcato-secund, broadly oblong lanceolate, hooked above, very concave, distinctly auricled at subdecurrent angles, glossy, lightly sulcate only when dry, nerve vanishing below apex, basal cells rectangular, narrower towards the margin, hyaline, yellowish; at angles rectangular and subquadrate, brownish-yellow. Caps. ovate-oblong, erect at base, arcuate, lid convex, shortly apiculate. June, July.

Bogs.

Scotland; near Birmingham, &c.

Var. β. WILSONI. St. very tall, sometimes 1 foot, yellow-green, with slender, nearly simple or irregularly pinnate branches. L. larger, with a filiform acumen, auricles very small.

Lancashire.

533. **H. exannulatum,** *Gümb.* (*aduncum,* Dill. L., ed. 1). St. 2—4 inches, erect, subpinnate. Br. short, simple, few. L. crowded, narrow, falcato-secund, lanceolate-

acuminate, striate, faintly subserrulate near the base, remotely denticulate above, nerved nearly to apex, basal cells oblong, rhomboid, gradually passing into the long, narrower ones above, at auriculate angles, larger, inflated, hyaline. Caps. subcylindrical, curved, cernuous, on a seta 1 inch long or more, lid convexo-conical, annulus absent.

Marshes and marshy heaths.   April, May.

Var. *β.* PURPURASCENS. Tufts deep, purple or variegated with green. L. except the youngest less elongate, purple, upper paler, less falcate.

Alpine pools.

Scotland.

Var. *γ.* STENOPHYLLUM, *Wils.* (*Rotæ*, De Not. et Sch. Syn.). Tufts purple, rarely green. St. densely and irregularly branched. L. long, straight, narrowly lanceolate, at apex of branches subsecund.

Subalpine streams and pools.

Yorkshire and North of England.

†† Branches and stems scarcely hooked.

534. **H. lycopodioides,** *Schw.* St. about 2 inches, erect, subpinnate, rather rigid. L. flexuoso-falcato-secund, ovate-acuminate, tapering to an acute point, but not apiculate, concave, flexuoso-sulcate, entire, nerved nearly to apex. Caps. oblong, cernuous, incurved from an erect base, lid conical, mamillate.

Bogs and marshes. Fruit rare.   June.
Southport, in fruit.

b. Monoicous.

535. **H. fluitans,** *Linn.* St. 6—12 inches, erect or floating, pinnate, slender. Br. short, deflexed; st. l.

flexuose, spreading, hooked at apex only; br. l. narrower, turned to one side, rarely falcato-secund, lanceolate, tapering from an ovate, rounded base, acuminate, slightly serrulate near apex, thinly nerved more than half way, not sulcate, areolæ enlarged at base. Caps. small, oblong, curved, subcernuous, on a very long seta, lid conical, mamillate.

Marshes, bogs, &c. April, May.

536. **H. revolvens**, *Swartz*. St. 2—4 inches, erect or procumbent, subpinnate. L. crowded, tortuous or circinnate, falcate, ovate-lanceolate, acuminate, concave, serrulate near apex, deep red or purplish, nerve strong, more than half way, areolæ not enlarged at base. Caps. oblong, cernuous, on a shorter seta, lid conical, acute.

Bogs and marshes. April, May.

537. **H. uncinatum**, *Hedw.* St. about 2 inches, slender, erect or procumbent, subpinnate. L. circinnate, secund, very narrow, lanceolate setaceous from a broader base, plicate, serrulate, nerved nearly to apex. Caps. cylindrical, curved, cernuous, lid conical.

Subalpine walls and rocks. May, June.

Sub-genus 4. *Cratoneuron.*

St. prostrate or ascending, radiculose and villose. Br. pinnate. Paraphyllia numerous, dissected. L. obcordate-lanceolate, falcato-secund, cells linear; at decurrent angles, lax, excavate, brown, opaque, *nerve thick.* Caps. oblong cylindric, on a long seta, cernuous from an erect neck. Dioicous.

538. **H. filicinum**, *Linn.* St. 2—4 inches, suberect, slender, pinnate, with purplish radicles. L. spreading, falcato-secund; st. l. deltoid-ovate, tapering; br. l.

ovate-lanceolate—all serrulate, scarcely twisted when
dry, nerved to or beyond apex, areolæ oval, rather
large, larger rhomboid and pellucid at base. Caps.
oblong, curved, cernuous, lid conical, acute. Dioicous.

Marshes, wet rocks.   April.

Var. β. VALLISCLAUSÆ, *Brid.* L. subsecund, nerve
very strong and excurrent.

Ormeshead; Derbyshire.

Var. γ. GRACILESCENS, *Schp.* Slender, pinnate, entirely prostrate, strongly tomentose. L. small, patent
or subsecund, deep green.

539. **H. commutatum,** *Hedw.* St. 2—4 inches or
more, procumbent; branches about ½ inch—both more
or less uncinate, radicles brownish; st. l. remote, from
a triangular, obcordate, bi-auriculate base, narrowly
falciform (upper auricle eroso-denticulate, lower decurrent, entire), tapering to a slender, long point, which
is twisted when dry, finely serrulate, nerved nearly
to apex; br. l. narrower—all much sulcate, areolæ
narrow, flexuose-linear; auricular, small, hexagono-
quadrate. Caps. large, oblong, lid conical. Dioicous.

Wet shady places.   April—July.

[*H. Breadalbanense,* F. B. White, seems to be only
a form of this species.]

540. **H. falcatum,** *Brid.* (*H. commutatum,* var. *con-
densatum,* Bry. Brit.). St. 2—3 inches, cæspitose,
erect, sparingly branched, scarcely radiculose. L.
larger, more or less falcate, ovate-oblong, lanceolate-
subulate, base minutely serrulate, less decurrent and
excavate, and less sulcate, nerved nearly to apex, which
is scarcely cirroso-flexuose. Caps. small, oval-oblong,
curved, cernuous.

Subalpine places and bogs.   May, June.

Var. β. GRACILESCENS. Irregularly branched, procumbent. L. smaller, shining, reddish-brown.

Yorkshire, Scotland.

Var. γ. VIRESCENS, *Boulay.* St. long. Br. fasciculate, pinnate. L. subfalcate-secund, apical, strongly falcate, scarcely auriculate, nerved close to apex, deep green, cells all narrow. Paraphyllia few. Probably a distinct species.

Malham, Yorkshire (W. West).

541. **H. sulcatum,** *Schpr.* Loosely cæspitose. St. rigid, without radicles, subpinnate. L. partly broadly elongate-lanceolate, partly sharply lanceolate from broadly ovate, minutely denticulate base, long and deeply sulcate, all reflexed, hamulose, nerve absent or slender, cells narrowly hexagono-rhomboid, obtuse at apex, broader at base, quadrate at auricles. Paraphyllia numerous.

Mountainous places.

Ben Lawers, July, 1865 (G. E. Hunt).

Sub-genus 5. *Rhytidium.*

Robust, prostrate or ascending, irregularly pinnate, without radicles. L. strongly *sulcato-rugose,* upper cells vermicular-linear, middle lower sinuoso-rectangular, at margin and angles minutely quadrate, nerve single, narrow. Calyptra large. Caps. oblong, cernuous or horizontal, lid beaked, annulus broad. Dioicous.

542. **H. rugosum,** *Ehr.* St. 2—3 inches, rigid, erect, densely tufted. Br. recurved; st. l. crowded, spirally imbricate and falcato-secund, serrulate and recurved at margin, lanceolate-acuminate from a broad base, rugose, nerved more than half way; br. l. often erecto-

patent.   Caps. subcylindrical, curved, pale reddish-brown, lid large, yellowish, with an oblique beak.

July.

Limestone and other rocks.   Barren in England.

Sub-genus 6.   *Homomallium.*

St. creeping, cæspitose, pinnate. L. *curved upwards, secund,* areolæ loosely rhomboid.   Caps. incurved, cernuous, compressed below mouth when dry.

543. **H. incurvatum,** *Schrad.*   St. short, slender. Branches curved upwards. L. ovate-lanceolate, tapering, all pointing upwards, entire or serrulate at apex, nerveless or shortly 2-nerved. Caps. small, ovate, horizontal, lid short, conical, acute. Monoicous.

Shady walls and stones.   June.

Sub-genus 7.   *Drepanium.*

St. more or less regularly pinnate. L. *falciform-secund,* two or singly nerved or nerveless, areolæ narrowly linear, quadrate at basal angles. Paraphyllia not numerous. Caps. subcylindrical, incurved, lid large, shortly rostellate.

a. Monoicous.

[544. **H. Canariense,** *Mitt.* (Journ. Linn. Soc. viii.). Tufted.   St. procumbent. Br. pinnate. L. falcato-secund, compressed, ovate-lanceolate or oval-lanceolate, gradually acute, margins sharply serrulate upwards, shortly 2-nerved, cells elongate, a few more obscure at the angles. Caps. shortly oval, with a large mouth, lid conical-acuminate. Dioicous. (Like *H. cupressiforme,* var. *mamillatus,* but differs in the sharp serrulation of the leaves, &c.)

"Killarney, 1829, Wilson."]

Derbyshire localities have been given, but the specimens belong to a form of *H. molluscum.*

545. **H. hamulosum,** *B. and S.* St. 1 inch or more, procumbent or ascending, radiculose, pinnate. Br. hooked at apex. L. circinnate-secund, much curved, tapering into a long, slender, subserrulate point, from an ovate-lanceolate base, nerveless or obsoletely 2-nerved, cells hexagono-vermicular, laxer at base, but scarcely excavate or dilated at angles. Caps. subcylindrical, curved, tapering at base, lid conical, pointed.

Alpine grassy declivities.   Summer.

Scotland, Ben Lawers, Craig Challeach, &c.

b. Dioicous.

546. **H. callichroum,** *Brid.* St. prostrate or ascending, slender, without radicles. Br. pinnate. L. crowded, falcate-secund and subcircinnate, from a broadly ovate, concave base, narrowly lanceolate, much acuminate, entire, almost or quite nerveless, cells narrow, sub-flexuose, at basal angles lax, excavate, and yellowish. Caps. oblong, incurved, cernuous, on a red, flexuose seta, lid orange, very convex, acutely beaked.

Damp stony mountainous places.   July, August.

North Wales, Scotland.

547. **H. imponens,** *Hedw.* Cæspitose, subpinnate. L. imbricate, circinnate secund, filiform from a broad, ovate-oblong base, margin reflexed below, minutely serrate, obsoletely 2-nerved, alar cells large, pellucid, orange-coloured; br. l. much narrower, and at apex of branches convolute and hamato-incurved; per. l. nerveless, filiform, flexuose, apiculate. Caps. suberect, cylindrical, incurved, lid convexo-conical, acutely pointed, yellowish, annulus broad.

Woods and stony ground.   Autumn.

Reigate Heath (Mr. Mitten), 1864; Strensall Common, York, &c.

548. **H. cupressiforme,** *Dill.* St. about 1 inch, procumbent, more or less regularly pinnate. L. falcato-secund, pointing downwards, sharply acuminate, from an ovate-lanceolate, auriculate base, entire or slightly serrulate at apex, nerveless or faintly 2-nerved; per. l. erect, almost piliferous. Caps. subcylindrical, cernuous, curved, lid conical, cuspidate.   Spring.

Walls, rocks, trunks of trees, &c.

Var. β. ERICETORUM (*compressum*). St. slender, pinnate, reddish, with compressed foliage. L. pale green, serrulate at apex, seta long, slender. Caps. short, elliptic, oblong. Amongst heath and bilberry, &c.

Var. γ. MINUS. Pinnate. Br. slender. L. narrow, falcate, serrulate, margin recurved. Caps. small, erect.

Trunks of trees.

Var. δ. FILIFORME. Br. prostrate, filiform, slender. L. falcate, serrulate. Caps. short, lid with a shorter point.

Rocks. Killarney, &c.

Var. ε. TECTORUM, *Schp.* (*lacunosum*, Wils.). More robust. Br. thickened. L. larger, subcoriaceous, yellowish-brown.

Var. θ. ELATUM. Robust, loosely cæspitose, erect. L. larger, broader, very concave, narrowly acuminate, brown or yellowish-green. Caps. erect, cylindrical, mouth incurved

Sutton Park, Birmingham (J. E. Bagnall).

549. **H. resupinatum,** *Wils.* St. creeping, subpinnate. L. erecto-patent, secund, pointing upwards, ovate-lanceolate, tapering to a point, entire, nerveless,

basal cells broad, excavate, orange. Caps. oblong, erect or cernuous, almost symmetrical, lid with an oblique beak.

Walls, rocks, trees, &c.   October, December.

550. **H. Patientiæ,** *Lindb.* (*Lindbergii,* Mitt., Journ. of Bot., i. p. 123. *II. pratense,* Bry. Brit., 399, but *non* Koch.). St. sparingly branched in an irregular manner, without any appearance of becoming pinnate. L. loosely compressed, ovate or ovate-lanceolate, acute, but with a broad point, concave. margins entire, nerveless, cells at angles enlarged and pale. Caps., according to Lindberg, is on a rather thick seta, 1 inch long, turgid, ovate, when dry arcuate, sulcate.

"Damp sandy ground among thin grass, not in bogs. The fruit has been gathered once by Dr. Klingraff in June, in W. Prussia."

(N.B.—The true *H. pratense* is not found in Britain, and *H. arcuatum,* Sull., is only found in the Sandwich and other islands in the Pacific Ocean.)

c. Fruit not known.

551. **H. Bambergeri,** *Schp.* Tufts rather small, dense, yellowish-green above, passing to yellow-fuscous at base. St. without radicles or villi, subpinnate. Br. few, fastigiate. L. densely crowded, secund, strongly circinnate, ovate-lanceolate, elongate, entire, with a long point, faintly nerved, single or unequally bifurcate, alar cells few, rather obscure, yellow, upper linear, elongate.

Near summit of Ben Lawers, July, 1867 (Dr. Fraser).

Sub-genus 8. *Ctenidium.*

Prostrate or ascending, rooting, regularly pinnate.

L. hamate, circinnato-secund. Caps. cernuous, short, solid, lid convexo-conical.

552. **H. molluscum,** *Hedw.* St. soft, 1—2 inches, suberect. L. circinnate secund; st. l. cordate; br. l. ovate-lanceolate—all tapering, acuminate, striate, serrulate, nerveless or faintly 2-nerved, crisped when dry. Paraphyllia ovate lanceolate. Caps. ovate, horizontal, lid conical, large, sharply pointed.   Summer.

Moist banks and limestone rocks. Common.

Var. FASTIGIATUM, *Boswell MS.* Branches erect.

This is the plant gathered in Derbyshire as *H. canariense.*

Sub-genus 9.   *Ctenium.*

St. erect, rigid, villose, regularly pinnate. Paraphyllia numerous. L. hamate. Caps. oblong, incurved, solid.

553. **H. Crista-castrensis,** *L.*   St. suberect, 3—4 in., pectinate; st. l. ovate-acuminate; br. l. narrowly lanceolate-acuminate, serrulate near apex—all circinnato-secund or tortuoso-falcate, deeply sulcate, faintly 2-nerved, margin reflexed. Caps. oblong, curved, cernuous, lid conical, pointed. Dioicous.   July, August.

Woods (chiefly pine) alpine and subalpine.

Sub-genus 10.   *Limnobium.*

St. soft, cæspitose, prostrate, irregularly branched. L. falcato-secund, rarely spreading, faintly nerved, areolæ linear; per. l. long, deeply sulcate. Caps. incurved, cernuous, lid convexo-conical or mamillate.

a. Monoicous.

554. **H. palustre,** *Linn.* St. creeping. Br. ascending,

crowded, curved, cuspidate and convolute at apex.   L. generally secund, sometimes almost falcate, ovate lanceolate, entire, strongly concave, pointed, either nerveless, shortly 2-nerved or singly nerved half way; per. l. erect, distinctly striate.   Caps. ovate or oblong, cylindric, slightly curved, cernuous, lid conical, pointed.

Stones and rocks in streams. May.

Var. β. L. imbricate, not secund.

Var. γ. SUBSPHŒRICARPON. L. strongly nerved nearly to apex. Caps. roundish ovate, tumid.

555. **H. molle,** *Dicks.* (*H. alpestre* (?), Bry. Eur., *non* Swartz). Very weak and flaccid, the tufts falling to pieces on removal from the water. L. varying from ovate to rotundo-ovate, flat, or sometimes very slightly reflexed towards apex, gradually tapering upwards, or very rarely suddenly apiculate, texture somewhat loose, areolæ larger and wider than in last, nerve rather long and thick, ill-defined, single or double. Caps. turgid, oval, incurved, cernuous, lid convex, mamillate.

Great elevations. August.

Ben-mac-Dhui, Ben Nevis.

556. **H. dilatatum,** *Wils.* (*H. molle,* Bry. Eur.). Plant of somewhat firm growth. L. rotundo-ovate, narrowly decurrent, at apex faintly serrate, rather concave, suddenly apiculate, texture very close, areolæ long and very narrow, slightly dilated at angles, hexagono-rectangular, nerve double, short, slender, but well defined. Caps. ovate, cernuous, curved, lid conical.

At a low elevation.

North Wales, Yorkshire, Berkshire, Clova, Braemar.

557. **H. arcticum,** *Sommerfelt.*   St. 1—2 inches, creeping, eradiculose. Br. elongate, simple, obtuse. L. spreading, green above, purplish below, small,

broadly ovate or roundish, somewhat obtuse, entire, plano-concave and subcochleariform, strongly 2-nerved about half way, sometimes nerves blended into one, cells narrowish, scarcely dilated at base. Caps. ovate, cernuous, tapering into the seta, lid mamillate.

Alpine rivulets.   June.

558. **H. eugyrium,** *Schp.* St. short, much branched. L. crowded; st. l. drooping on two sides, broadly oblong-lanceolate, shortly acuminate; br. l. flexuoso-falcate, plano-concave, elongate-lanceolate, narrower, serrulate at apex, nerve thin, unequally bifid, areolæ vermicular; excavate, large, fulvous and rectangular at the decurrent angles; per. l. external spreading, internal erect, longly lanceolate, with erose apices. Caps. ovate-oblong, cernuous, turgid, lid mammillate. Annulus broadly bi-triseriate.

Stones in waterfalls.   Summer.
North Wales, Devonshire, Killarney.

Var. β. MACKAYI, *Schp.* Denser and more robust. L. subsecund or erecto-patent, broadly oblong, less acuminate; per. l. shorter, int. patent, less sulcate. Annulus narrower.

Killarney, Kendal, Fife, Bangor, Dent (Yorkshire), Devonshire (auct. Schp.).

b. Dioicous.

559. **H. ochraceum,** *Turn.* St. 2—4 inches, tufted, filiform, suberect, sparingly branched, ochraceous below, eradiculose. L. yellowish-green, subsecund, sometimes falcate, distant, ovate-lanceolate, pointed, concave, sulcate, nerve forked, extending half way, entire or faintly serrate at apex, cells narrowly flexuose, at angles lax, rectangular, hyaline; per. l. squarrose,

recurved. Caps. oblong, tapering at base, cernuous, lid conical.   May, June.

Stones in alpine and subalpine streams, &c.

Var. β. FLACCIDUM, *Milde.* Yorkshire.

Sub-genus 11. *Chrysobryum.*

Prostrate, slender, flaccid, forming flat, golden-glossy patches. L. roundish or elliptical, concave, nerveless or faintly 2-nerved, cells largish, linear, fusiform, sometimes enlarged hyaline at angles.

560. **H. micans,** *Wils.* St. prostrate, very slender, filiform, branched. L. small, subsecund, almost orbicular, concave, apiculate, serrulate, sometimes faintly 2-nerved at base; br. l. smaller. Fruit not known.

Shady rocks.

South of Ireland, Borrowdale.

Sub-genus 12. *Calliergon.*

Erect or procumbent, stem simple or more or less pinnate, generally eradiculose. L. patent, rarely subcomplanate or subsecund, cordate-ovate and ovate-oblong, thinly single-nerved or shortly 2-nerved, shining, areolæ linear. Caps. oblong, incurved, cernuous, lid mamillate or convexo-conical.

a. Stem more or less regularly pinnate. Leaves patent or loosely imbricate.

Monoicous.

561. **H. cordifolium,** *Hedw.* Bright green above, reddish-brown below. St. 3—6 in., erect, subpinnate. Br. short, slender. L. spreading, almost squarrose, convolute and cuspidate at tip of branches, distant, cordate-ovate, obtuse or slightly apiculate, concave,

entire, strongly nerved almost to apex, cells narrow above, at middle lax rhomboid, hexagono-rectangular at base and angles. Caps. oblong, cylindric, incurved, horizontal, lid mamillate.

Marshes and ditches.   April, May.

Dioicous.

562. **H. giganteum,** *Schp.* St. erect, thick, often 1 foot long, densely pinnate; st. l. patent, broadly cordate-ovate, strongly nerved to often cucullate apex, cells linear, excavate, quadrate, and hyaline or rufous-brown at basal angles; br. l. lingulate, narrow, terminal ones twisted and subulate; per. l. oblong-lanceolate. Caps. oblong-cylindrical, subincurved, horizontal, on a long seta. Annulus none, lid mamillate.

Marshes.   Summer.

Hale Moss and Wybunbury Bog, &c.

563. **H. sarmentosum,** *Wahl.* St. 1 inch or more, procumbent, subpinnate, without radicles. Br. short, cuspidate, young foliage green, the rest red or purplish. L. much crowded, suberect, elliptic-oblong, scarcely pointed, concave, entire, nerved almost to apex, which is subcucullate, areolæ large, quadrate and pellucid at basal angles. Caps. ovate-oblong, cernuous or horizontal, lid mamillate.

Wet alpine rocks.   Summer.

Var. β. SUBFLAVUM, *Ferg.* Scotland.

564. **H. cuspidatum,** *Linn.* St. 2—6 inches, erect, pinnate, terminal foliage cuspidate. L. spreading, almost squarrose, when young erect, appressed, and convolute, broadly ovate-oblong, obtuse, concave, entire, nerveless or shortly 2-nerved, margins more or less incurved at apex, cells enlarged and pellucid at

decurrent basal angles. Caps. oblong, much curved, tapering below, lid conical, acute.

Marshes. May, June.

Var. β. PUNGENS. St. and br. fragile. Br. circinnate. L. nerveless; st. l. imbricate, less decurrent; br. l. all strongly convolute (so that the branches are terete, subuliform), colour pale olive.

Yorkshire, Scotland.

565. **H. Schreberi,** *Willd.* St. 4—6 inches, erect, pinnate, deep red, with slender, curved branches, somewhat cuspidate at summit. L. convolute, afterwards erecto-patent, elliptical or ovate-oblong, concave, obtuse or slightly acuminate, subsulcate, margin recurved at base, incurved above, shortly 2-nerved, cells at basal angles enlarged, quadrate, orange. Caps. ovate-oblong, curved, cernuous, lid conical, pointed.

Woods and shady banks. Autumn.

566. **H. purum,** *Linn.* St. 4—6 inches, not coloured, erect, pinnate. Br. slightly curved, not cuspidate at apex. L. turgid, imbricate, broadly elliptical, concave, with recurved points, almost boat-shaped, entire, nerved half way, minutely serrulate; br. l. smaller. Caps. ovate, suddenly horizontal, lid conical.

Shady banks. Spring.

b. Stem almost simple or sparingly branched. Leaves closely imbricate when dry.

Dioicous.

567. **H. stramineum,** *Dicks.* St. 2—4 inches, erect, filiform, with few erect branches. L. erecto-patent, elliptic-oblong, obtuse, entire, concave, thinly nerved nearly to apex, cells enlarged, quadrate, and pellucid

at basal angles.   Caps. small, ovate, curved, cernuous,
lid short, conical.   April, May.

Marshes amongst *Sphagnum.*   Rare in fruit.

568. **H. trifarium,** *W. and M.*   St. 2—3 inches, erect
or trailing, sparingly branched.   L. very closely imbricate, fragile when dry, generally but not always
trifarious, roundish obtuse, slightly decurrent, concave,
inflated, entire, nerved almost to apex or shortly
2-nerved. · Caps. oval-oblong, curved, cernuous, lid
conical.

Alpine bogs and turfy rills.   June (?).

Sub-genus 13.   *Scorpidium.*

Fastigiate, sparingly branched.   Leaves imbricate,
secund, ventricose, faintly nerved, areolæ narrow.

Dioicous.

569. **H. scorpioides,** *Linn.*   St. 3—4 inches, erect or
procumbent, irregularly pinnate.   Branches short.   L.
crowded, imbricate, falcato-secund, large, roundish-
ovate, ventricose, apiculate, entire, nerveless or faintly
and shortly 2-nerved, purplish-brown or lurid.   Caps.
short, oblong, curved, tumid, cernuous, on a long seta,
lid conical, pointed.

Bogs.   May.

129.   **HYLOCOMIUM,** *Sch.*

a. Regularly bi-tripinnate.   Leaves loosely imbricate,
lid rostrate.

570. **H. splendens,** *Hedw.*   St. 2—6 inches, erect or
procumbent, interruptedly bi-tripinnate, villous, reddish ; foliage reddish or fulvous-green ; st. l. roundish-
elliptical, with long, wavy points ; br. l. with a short
point or muticous—all imbricate, concave, serrate,

shortly 2-nerved, margin recurved below. Caps. ovate, curved, cernuous, lid convex, tapering into a long beak. Dioicous.

Grassy banks, woods, &c. April.

b. Irregularly pinnate. Leaves patent, lid mamillate or shortly beaked.

571. **H. umbratum,** *Ehrh.* St. arched, suberect, with branched villi, irregularly bipinnate. L. yellowish-green, glossy, decurrent, obcordate-acuminate, serrate, plicato-striate, nerve unequally bifurcate ; br. l. smaller, ovate, and less acuminate. Caps. short, roundish, obovate, curved, cernuous, lid conical, acute. Dioicous.

Alpine woods on stones. November.

572. **H. Oakesii,** *Sulliv.* St. arched, irregularly and distantly pinnate, with branched villi. L. larger, elliptical, concave, not cordate, plicato-striate, sharply acuminate, from middle upwards serrate or coarsely denticulate, singly nerved half way or shortly 2-nerved, margin recurved ; per. l. squarrose. Caps. roundish ovate, turgid, cernuous, lid conical, shortly beaked. Dioicous.

Alpine rocks. Autumn (?).

573. **H. brevirostrum,** *Ehrh.* St. 2—6 inches, arched, erect, with branched villi, irregularly bipinnate ; st. l. distant, almost squarrose, plicato-striate, cordate, and suddenly acuminate ; br. l. ovate-acuminate, not so suddenly acuminate, striate—all serrulate and 2-nerved half way. Caps. roundish ovate, cernuous, lid conical, tapering into a rather long, inclined beak. Dioicous.

Mountainous woods. Spring.

c. Leaves squarrose.

574. **H. squarrosum,** *Linn.* St. 2—3 inches, reddish,

slender, more or less erect, irregularly pinnate. Br.
drooping; st. l. squarrose, recurved, ovate, gradually
tapering and very acute, faintly striate below; br. l.
narrower, less recurved and squarrose—all obsoletely
serrulate and shortly 2-nerved. Caps. roundish ovate,
drooping, lid conical, with a short, sharp point.
Dioicous.

Banks and woods.   November.

Var. CALVESCENS, *Wils.* Br. more or less regularly
pinnate. St. l. slightly broader, subsulcate, distinctly
serrate at apex.

575. **H. triquetrum,** *Linn.* St. 6 inches or more,
rigid, reddish, erect, subpinnate. Br. long, straggling;
st. l. squarrose or subsecund, striate; br. l. spreading,
scarcely striate—all triangular acuminate, from a cordate, amplexicaul base, serrulate and 2-nerved half
way. Caps. roundish ovate, cernuous, lid conical,
acute. Dioicous.

Woods, &c.   Spring.

d. Leaves secund or falcato-secund.

576. **H. loreum,** *Linn.* St. 6—12 inches, slender,
erect or procumbent, more or less pinnate. Br. drooping, straggling. L. squarrose, recurved, more or less
secund at summit of stem and branches, ovate-lanceolate, with a long acumen, not cordate or amplexicaul,
plicato-striate below, shortly and faintly 2-nerved,
sometimes nerveless. Caps. small, roundish ovate, lid
conical, sharply pointed. Dioicous.

Mountainous woods.   November.

GLOSSARY

PRINCIPAL TERMS USED IN THIS VOLUME.

Acuminate, taper-pointed.

Acute, pointed; scarcely tapering.

Alar (cells), at basal angles.

Annulus, an elastic ring round mouth of capsule.

Apicu-lus (-late), a very short point.

Apophysis, an excrescence; a swelling at base of capsule.

Arcuate, arched or curved.

Areolæ, the leaf cells.

Arista, a short bristly point.

Auricles (of leaf), short lobes on each side of base.

Bifarious, two-ranked.

Cæspitose, tufted or matted together.

Calyptra, the outermost covering, or veil, of the capsule.

Capsule, the fruit, enclosing the spores.

Carinate, keeled.

Cernuous, nodding.

Chlorophyll, the green matter filling the cells.

Cilia, hair-like divisions of the inner peristome.

Circinnate, curved nearly into a circle.

Cirrhose, having a very narrow hair-like wavy point.

Clathrate, trellised.

Clavate, club-shaped.

Cochleariform, spoon-shaped.

Columella, the central pillar of capsule round which the spores are grouped.

Comal, the large topmost leaves of some stems.

Complanate, flat.

Convoluta, rolled together.
Cribrose, like a sieve.
Crura (legs), referring to teeth of peristome.
Cucullate (*cucullus*), hooded.
Cuspidate, with a short bristly point.
Cyathiform, cup-shaped.
Decurrent (of leaf), running down the stem.
Dendroid, tree-like.
Dentate, toothed.
Denticulate, with smaller teeth.
Diaphanous, semi-transparent.
Dichotomous, forked.
Dimidiate, split up one side.
Dioicous, barren and fertile flowers on different plants.
Distichous, inserted in two opposite rows.
Divaricate, widely spreading.
Dorsal, at the back.
Erose, as if bitten or gnawed out.
Excurrent (of nerve), continued beyond the apex of leaf.
Exserted, standing out from the leaves.
Falcate, falchion-shaped, or much bent.
Fasciculate (stems or branches), of unequal height.
Fastigiate   ditto   reaching to same height.
Filiform, thread-like.
Fugacious, falling early.
Gemmiform or *Gemmaceous*, like a bud.
Geniculate, suddenly bent like the leg when kneeling.
Gibbous, swelling out, protuberant.
Granulate, roughly dotted on surface.
Gregarious, growing together, but not matted.
Hamate, Hamulose, bent like a hook.
Hyaline, glassy.
Hygrometric (*hygroscopic*), moving when moistened.
Imbricate, overlapping each other like tiles.
Immersed (of capsule), when almost buried in the leaves.
Inflexed, bent inwards.
Julaceous, resembling a slender glossy worm.
Lamina, the blade of the leaf.
Lid, the cover to the mouth of capsule.

Ligulate, strap-shaped.

Lingulate, tongue-shaped.

Mitriform (of calyptra), mitre-shaped, not split up the side.

Monoicous, barren and fertile flowers on same plant, but not on same receptacle.

Mucro, a short terminal point.

Mucronate, terminated with a mucro.

Muriculate, roughened with sharpish prominences.

Muticous, without a point.

Ochrea, the filmy sheath surrounding base of seta.

Pagina, the blade of the leaf apart from the nerve.

Panduriform, fiddle-shaped.

Papillose, roughened with blunt roundish prominences.

Paraphyllia, branched or unbranched processes between the leaves.

Percurrent, extending the entire length.

Perichætium, the leafy involucre at base of seta, surrounding the vaginula in fertile flowers.

Perigonium, the leaves surrounding the barren flowers.

Peristome, the teeth at mouth of capsule covered by the lid before it falls.

Plicate, furrowed.

Præmorse, ending suddenly, as if bitten off.

Pyriform, pear-shaped.

Quadrate, square.

Radicles, small rooting fibres.

Radiculose, covered with radicles.

Rhizome, a creeping subterranean stem.

Rhomboid (*rhombus*), an oblique square.

Rostellate, with a very short beak.

Rostrate, with a longer beak.

Rugose, wrinkled or crumpled.

Scariose, dry and chaffy (opposed to tender and succulent).

Secund, all turned to one side.

Seta, the fruit-stalk.

Setaceous, bristle-shaped.

Sigmatoid, bent like the letter S.

Spathulate, somewhat resembling a battledore.
Strumose, swollen at base.
Sub-, in a slight degree; e.g. "sub-serrate," slightly
  serrate.
Subula, an awl.
Subulate, awl-shaped.
Sulcate, furrowed.
Synoicous, male and female flowers on same receptacle.
Terete, cylindrical.
Thœca, the capsule.
Tomentose, covered with down.
Trabeculate, barred.
Tristichous, three-ranked.
Truncate, having the point cut off.
Tumid, swollen.
Turbinate, shaped like a peg-top.
Uncinate, bent like a hook.
Undulate, wavy.
Vaginula, the cellular sheath surrounding the base of
  the seta.
Vermicular, narrow and wavy (like a worm).
Villi, short leafy processes on the stem amongst the
  leaves.
Villous, covered with villi.

INDEX .

TO GENERA AND THEIR SYNONYMS.

The Books in this Catalogue have been reduced to net cash prices, and are sent Post-free on receipt of remittance. All previous Catalogues are withdrawn.

LIST OF WORKS

ON

NATURAL HISTORY, TOPOGRAPHY, ANTIQUITY, AND SCIENCE.

CONTENTS.

PUBLISHED BY

LOVELL REEVE & CO., Limited,

PUBLISHERS TO THE HOME, COLONIAL, AND INDIAN GOVERNMENTS,

6, HENRIETTA STREET, COVENT GARDEN, W.C.

LOVELL REEVE & CO.'S
Crown Series of Natural History.

For descriptive details, see Catalogue.

These handy and well-illustrated Volumes, while popular in style to suit beginners, are strictly scientific in method, and form excellent introductions to more advanced works. They are admirably adapted for school prizes and presents. .

Handbook of the British Flora. By G. BENTHAM, F.R.S. Revised by Sir J. D. HOOKER, C.B., G.C.S.I., F.R.S., &c. 9s.

Illustrations of the British Flora. Drawn by W. H. FITCH, F.L.S., and W. G. SMITH, F.L.S. 1315 Wood Engravings. 7th Edition, revised and enlarged, 9s.

British Grasses. By M. PLUES. 16 Coloured Plates, and Woodcuts, 9s.

British Ferns. By M. PLUES. 16 Coloured Plates, and Woodcuts, 9s.

British Seaweeds. By S. O. GRAY. 16 Coloured Plates, 9s.

Synopsis of British Mosses. By C. P. HOBKIRK, F.L.S. Revised Edition, 6s. 6d.

British Insects. By E. F. STAVELEY. 16 Coloured Plates, and Woodcuts, 12s.

British Beetles. By E. C. RYE. 2nd Edition, revised by Rev. CANON FOWLER, M.A., F.L.S. 16 Coloured Plates, and Woodcuts, 9s.

British Butterflies and Moths. By H. T. STAINTON. 2nd Edition, 16 Coloured Plates, and Woodcuts, 9s.

British Bees. By W. E. SHUCKARD. 16 Coloured Plates, and Woodcuts, 9s.

British Spiders. By E. F. STAVELEY. 16 Coloured Plates, and Woodcuts, 9s.

British Zoophytes. By ARTHUR S. PENNINGTON, F.L.S. 24 Plates, 9s.

The Edible Mollusca of Great Britain and Ireland, with Recipes for Cooking them. By M. S. LOVELL. Second Edition. 12 Coloured Plates, 9s.

BOTANY.

The Botanical Magazine; Figures and Descriptions

of New and Rare Plants suitable for the Garden, Stove, or Green-house. Fourth Series. Edited by D. PRAIN, C.I.E., LL.D., F.R.S., Director of the Royal Gardens, Kew. Vols. I.—V., Royal 8vo, 42s. each. Published Monthly, with 6 Plates, 3s. 6d., coloured. Annual Subscription, 42s.

COMPLETION of the THIRD SERIES in 60 Vols., with nearly 4000 Coloured Plates, 42s. each; to Subscribers for the entire Series, 36s. each.

Monographs from the Third Series of the Botanical

Magazine, in which the plates and descriptions illustrating each Genus are brought together under one wrapper. A List may be had on application. The prices vary from 1s. to 51s.

A New and Complete Index to the Botanical

Magazine. Vols. I. to CXXX. Comprising the 1st, 2nd, and 3rd Series. To which is prefixed a History of the Magazine by W. BOTTING HEMSLEY. 21s.

Catalogue of the Plants of Kumaon and of the

adjacent portions of Garhwal and Tibet. By Lt.-Gen. Sir RICHARD STRACHEY and J. F. DUTHIE. 5s.

The Uses of British Plants. Traced from

antiquity to the present day, together with the derivations of their names. By the Rev. Prof. G. HENSLOW, M.A., F.L.S. With 288 Illustrations. Crown 8vo, 4s. 6d.

Handbook of the British Flora; a Description of

the Flowering Plants and Ferns indigenous to, or naturalized in, the British Isles. For the use of Beginners and Amateurs. By GEORGE BENTHAM, F.R.S. Revised by Sir J. D. HOOKER, C.B., G.C.S.I., F.R.S. Crown 8vo, 9s.

Illustrations of the British Flora; a Series of

Wood Engravings, with Dissections, of British Plants, from Drawings by W. H. FITCH, F.L.S., and W. G. SMITH, F.L.S., forming an Illustrated Companion to BENTHAM's "Handbook," and other British Floras. 7th Edition, revised and enlarged. 1315 Wood Engravings, 9s.

Outlines of Elementary Botany, as Introductory
to Local Floras. By GEORGE BENTHAM, F.R.S., F.L.S. New Edition, 1s.

The Narcissus, its History and Culture, with
Coloured Figures of all known Species and Principal Varieties. By F. W. BURBIDGE, and a Review of the Classification by J. G. BAKER, F.L.S. Super-royal 8vo, 48 Coloured Plates, 30s.

The Natural History of Plants. By H. BAILLON,
President of the Linnæan Society of Paris, Professor of Medical Natural History and Director of the Botanical Garden of the Faculty of Medicine of Paris. Super-royal 8vo. Vols. I. to VIII., with 3545 Wood Engravings, 21s. each.

The Floral Magazine; New Series, Enlarged to
Royal 4to. Figures and Descriptions of the choicest New Flowers for the Garden, Stove, or Conservatory. Complete in Ten Vols., in handsome cloth, gilt edges, 36s. each.

FIRST SERIES complete in Ten Vols., with 560 beautifully-coloured Plates, £15 15s.

The Young Collector's Handybook of Botany.
By the Rev. H. P. DUNSTER, M.A. 66 Woodcuts, 3s.

Materials for a Flora of the Malayan Peninsula.
By H. N. RIDLEY, M.A., F.R.S., Director of Botanic Gardens, Singapore. Complete in Three Parts, 30s.

British Wild Flowers, Familiarly Described in
the Four Seasons. By THOMAS MOORE, F.L.S. 24 Coloured Plates. 14s.

Flora Vitiensis; a Description of the Plants of
the Viti or Fiji Islands, with an Account of their History, Uses, and Properties. By Dr. BERTHOLD SEEMANN, F.L.S. Royal 4to, Coloured Plates. Part X., 25s.

Flora Hongkongensis; a Description of the
Flowering Plants and Ferns of the Island of Hongkong. By GEORGE BENTHAM, F.R.S. With a Supplement by Dr. HANCE. 21s. Published under the authority of the Secretary of State for the Colonies. The Supplement separately, 2s. 6d.

Flora of Mauritius and the Seychelles; a Description of the Flowering Plants and Ferns of those Islands. By J. G. BAKER, F.L.S. 24s. Published under the authority of the Colonial Government of Mauritius.

Flora of British India. By Sir J. D. HOOKER, G.C.S.I., C.B., F.R.S., &c.; assisted by various Botanists. Complete in Seven Vols., cloth, £12. Published under the authority of the Secretary of State for India in Council.

Flora of Tropical Africa. By DANIEL OLIVER, F.R.S., F.L.S. Vols. I. to III., 20s. each. Continuation. Edited by Sir W. T. THISELTON-DYER, F.R.S., F.L.S. Vol. IV., Section 1, 30s. Section 2, 27s. Vol. V., 25s. 6d. Vol. VI., Section 1, Part 1, 8s. Vol. VII., 27s. 6d. Vol. VIII., 25s. 6d. Published under the authority of the Secretary of State for the Colonies.

Handbook of the New Zealand Flora; a Systematic Description of the Native Plants of New Zealand, and the Chatham, Kermadec's, Lord Auckland's, Campbell's, and Macquarrie's Islands. By Sir J. D. HOOKER, G.C.S.I., F.R.S. 42s. Published under the auspices of the Government of that Colony.

Flora Australiensis; a Description of the Plants of the Australian Territory. By GEORGE BENTHAM, F.R.S., assisted by FERDINAND MUELLER, F.R.S., Government Botanist, Melbourne, Victoria. Complete in Seven Vols., £7 4s. Published under the auspices of the several Governments of Australia.

Flora of the British West Indian Islands. By Dr. GRISEBACH, F.L.S. 42s. Published under the auspices of the Secretary of State for the Colonies.

Flora Capensis; a Systematic Description of the Plants of the Cape Colony, Caffraria, and Port Natal. By W. H. HARVEY, M.D., F.R.S., and O. W. SONDER, Ph.D. Vols. I. to III., 20s. each. Continuation. Edited by Sir W. T. THISELTON-DYER, C.M.G., C.I.E., LL.D., F.R.S. Vol. IV., Section 1, 52s. Section 2, 24s. Vol. V., Part I., 9s. Vol. VI., 24s. Vol. VII., 33s.

Genera Plantarum, ad Exemplaria imprimis in
Herbariis Kewensibus servata definita. By GEORGE BENTHAM,
F.R.S., F.L.S., and Sir J. D. HOOKER, F.R.S., late Director of the
Royal Gardens, Kew. Complete in 7 Parts, forming 3 Vols., £8 2s.

Flora of West Yorkshire; with an Account of the
Climatology and Lithology in connection therewith. By FREDERIC
ARNOLD LEES, M.R.C.S. Eng., L.R.C.P. Lond., Recorder for the
Botanical Record Club, and President of the Botanical Section of
the Yorkshire Naturalists' Union. With Coloured Map, 21s.

Flora of Hampshire, including the Isle of Wight,
with localities of the less common species. By F. TOWNSEND,
M.A., F.L.S. 2nd Edition, greatly enlarged and improved.
With large Coloured Map and two Plates, demy 8vo, 21s.

British Grasses; an Introduction to the Study
of the Gramineæ of Great Britain and Ireland. By M.
PLUES. Crown 8vo, with 16 Coloured Plates and 100 Wood
Engravings, 9s.

Insular Floras. A Lecture delivered by Sir J. D.
HOOKER, C.B., before the British Association for the advance-
ment of Science, at Nottingham, August 27, 1866. 2s. 6d.

Icones Plantarum. Figures, with Brief Descrip-
tive Characters and Remarks, of New and Rare Plants, selected
from the Author's Herbarium. By Sir W. J. HOOKER, F.R.S.
New Series, Vol. V. 100 Plates, 31s. 6d.

Botanical Names for English Readers. By RANDAL
H. ALCOCK. 8vo, 6s.

A Second Century of Orchidaceous Plants, selected
from the Subjects published in Curtis's "Botanical Magazine"
since the issue of the "First Century." Edited by JAMES BATE-
MAN, Esq., F.R.S. Complete in One Vol., Royal 4to, 100 Coloured
Plates, £5 5s.

*Dedicated by Special Permission to H.R.H. the Princess of Wales,
now H.M. Queen Alexandra.*

Monograph of Odontoglossum, a Genus of the
Vandeous Section of Orchidaceous Plants. By JAMES BATEMAN,
Esq., F.R.S. Imperial folio, in One Vol., with 30 Coloured Plates,
and Wood Engravings, cloth, £6 16s. 6d.

The Rhododendrons of Sikkim-Himalaya; being
an Account, Botanical and Geographical, of the Rhododendrons
recently discovered in the Mountains of Eastern Himalaya by
Sir J. D. Hooker, F.R.S. By Sir W. J. HOOKER, F.R.S. Folio,
30 Coloured Plates, £4 14s. 6d.

The Potamogetons of the British Isles; De-
scriptions of all the Species, Varieties, and Hybrids. By
ALFRED FRYER, A.L.S., Illustrated by ROBERT MORGAN, F.L.S.
Royal 4to. Sections 1, 2 and 3, containing parts 1-3, 4-6, 7-9,
each with 12 Plates, 21s. coloured; 15s. uncoloured.

FERNS.

British Ferns; an Introduction to the Study of
the FERNS, LYCOPODS, and EQUISETA indigenous to the British
Isles. With Chapters on the Structure, Propagation, Cultivation,
Diseases, Uses, Preservation, and Distribution of Ferns. By
M. PLUES. Crown 8vo, with 16 Coloured Plates, and 55 Wood
Engravings, 9s.

The British Ferns; Coloured Figures and Descrip-
tions, with Analysis of the Fructification and Venation of the
Ferns of Great Britain and Ireland. By Sir W. J. HOOKER,
F.R.S. Royal 8vo, 66 Coloured Plates, 36s.

Garden Ferns; Coloured Figures and Descriptions,
with Analysis of the Fructification and Venation of a Selection of
Exotic Ferns, adapted for Cultivation in the Garden, Hothouse,
and Conservatory. By Sir W. J. HOOKER, F.R.S. Royal 8vo,
64 Coloured Plates, 36s.

Filices Exoticæ; Coloured Figures and Description
of Exotic Ferns. By Sir W. J. HOOKER, F.R.S. Royal 4to,
100 Coloured Plates, £6 11s.

Ferny Combes; a Ramble after Ferns in the Glens
and Valleys of Devonshire. By CHARLOTTE CHANTER. Third
Edition. Fcap. 8vo, 8 Coloured Plates and a Map of the
County, 3s. 6d.

MOSSES AND HEPATICÆ.

Synopsis of British Mosses, containing Description of all the Genera and Species (with localities of the rarer ones) found in Great Britain and Ireland. By CHARLES P. HOBKIRK, F.L.S., &c., &c. New Edition, entirely revised. Crown 8vo, 6s. 6d.

Handbook of British Mosses, containing all that are known to be natives of the British Isles. By the Rev. M. J. BERKELEY, M.A., F.L.S. Second Edition. 24 Coloured Plates, 21s.

The Hepaticæ of the British Isles; being Figures and Descriptions of all known British Species. By WILLIAM HENRY PEARSON. New and cheaper issue. Complete in 2 Vols., with 228 Plates, plain, £5 5s.; coloured, £7 10s.

FUNGI.

British Fungi, Phycomycetes and Ustilagineæ. By GEORGE MASSEE (Lecturer on Botany to the London Society for the Extension of University Teaching). Crown 8vo, with 8 Plates, 6s. 6d.

Outlines of British Fungology. By the Rev. M. J. BERKELEY, M.A., F.L.S. With a Supplement of nearly 400 pages by WORTHINGTON G. SMITH, F.L.S., bringing the work down to the present state of Science. Two vols., 24 Coloured Plates, 36s. The SUPPLEMENT separately, 12s.

The Esculent Funguses of England. Containing an Account of their Classical History, Uses, Characters, Development, Structure, Nutritious Properties, Modes of Cooking and Preserving, &c. By C. D. BADHAM, M.D. Second Edition. Edited by F. CURREY, F.R.S. 12 Coloured Plates, 12s.

ALGÆ.

British Seaweeds; an Introduction to the Study of
the Marine Algæ of Great Britain, Ireland, and the Channel Islands.
By S. O. Gray. Crown 8vo, with 16 Coloured Plates, 9s.

Phycologia Britannica; or, History of British
Seaweeds. Containing Coloured Figures, Generic and Specific
Characters, Synonyms and Descriptions of all the Species of Algæ
inhabiting the Shores of the British Islands. By Dr. W. H.
Harvey, F.R.S. New Edition. Royal 8vo, 4 vols. 360
Coloured Plates, £6 6s.

Phycologia Australica; a History of Australian
Seaweeds, comprising Coloured Figures and Descriptions of the
more characteristic Marine Algæ of New South Wales, Victoria,
Tasmania, South Australia, and Western Australia, and a
Synopsis of all known Australian Algæ. By Dr. W. H. Harvey,
F.R.S. Royal 8vo, Five Vols., 300 Coloured Plates, £6 10s.

SHELLS AND MOLLUSKS.

Elements of Conchology; an Introduction to
the Natural History of Shells, and of the Animals which form
them. By Lovell Reeve, F.L.S. Royal 8vo, Two Vols., 62
Coloured Plates, £2 16s.

Conchologia Iconica; or, Figures and Descriptions
of the Shells of Mollusks, with remarks on their Affinities, Syno-
nymy, and Geographical Distribution. By Lovell Reeve,
F.L.S., and G. B. Sowerby, F.L.S. Complete in Twenty Vols.,
4to, with 2727 Coloured Plates, half-calf, £178.
A detailed list of Monographs and Volumes may be had.

The Edible Mollusca of Great Britain and Ireland,
with the Modes of Cooking them. By M. S. Lovell. With
12 Coloured Plates. New Edition, rewritten and much enlarged, 9s.

Testacea Atlantica; or, the Land and Freshwater
Shells of the Azores, Madeiras, Salvages, Canaries, Cape Verdes,
and Saint Helena. By T. Vernon Wollaston, M.A., F.L.S.
Demy 8vo, 21s.

ENTOMOLOGY.

A Monograph of the Genus Teracolus. By
E. M. BOWDLER SHARPE. Parts 1 to 11, 4to, each with 4 Coloured Plates, 7s. 6d.

A Monograph of the Membracidæ. By GEORGE
BOWDLER BUCKTON, F.R.S., F.L.S. To which is added a Paper entitled "Suggestions as to the Meaning of the Shapes and Colours of the Membracidæ in the Struggle for Existence," by EDWARD B. POULTON, D.Sc., M.A., Hon. LL.D. (Princeton), F.R.S., &c., Hope Professor of Zoology in the University of Oxford. Complete in One Vol. 4to, with 2 Structural and 60 Coloured Plates, cloth, gilt tops, £6 15s.

Monograph of the British Cicadæ or Tettigidæ.
By GEORGE BOWDLER BUCKTON, F.R.S., F.L.S., F.C.S., F.E.S., &c. Two Vols. 8vo, 82 Coloured Plates, 42s.

The Natural History of Eristalis Tenax, or the
Drone-Fly. By GEORGE BOWDLER BUCKTON, F.R.S., F.L.S., &c. 9 Plates, some Coloured, 8s.

The Hymenoptera Aculeata of the British Isles.
By EDWARD SAUNDERS, F.L.S. Complete in One Vol., with 3 Structural Plates, 16s. Large Paper Edition, with 51 Coloured Plates, 68s.

The Hemiptera Heteroptera of the British Islands.
By EDWARD SAUNDERS, F.L.S. Complete in One Vol., with a Structural Plate, 14s. Large Paper Illustrated Edition, with 31 Coloured Plates, 48s.

The Hemiptera Homoptera of the British Islands.
A Descriptive Account of the Families, Genera, and Species indigenous to Great Britain and Ireland, with Notes as to Localities, Habitats, &c. By JAMES EDWARDS, F.E.S. Complete in One Vol., with 2 Structural Plates, 12s. Large Paper, with 28 Coloured Plates, 43s.

Dedicated, by Special Permission, to Her late Majesty Queen Victoria, Empress of India.

Lepidoptera Indica. By F. MOORE. The Con-
tinuation by Col. C. SWINHOE. 4to. Vol. I., with 94, and Vols. II. to V., with 96, Coloured Plates; Vol. VI., with 84 Coloured Plates; each £9 5s., cloth. Parts 73—80, 15s. each.

The Lepidoptera of Ceylon. By F. MOORE, F.L.S.
Three Vols., Medium 4to, 215 Coloured Plates, cloth, gilt tops, £21 12s. Published under the auspices of the Government of Ceylon.

The Lepidoptera of the British Islands. By
CHARLES G. BARRETT, F.E.S. Complete in 11 vols, £6 12s. Large Paper Edition, with 504 Coloured Plates, £33 15s. Alphabetical List of Species contained in the work, 1s. 6d. Large Paper Edition, 2s.

Labelling List of the British Macro-Lepidoptera,
as arranged in "Lepidoptera of the British Islands." By CHARLES G. BARRETT, F.E.S. 1s. 6d.

The Larvæ of the British Lepidoptera, and their
Food Plants. By OWEN S. WILSON. With Life-sized Figures drawn and coloured from Nature, by ELEANORA WILSON. Super-royal 8vo, with 40 Coloured Plates. 63s.

The Coleoptera of the British Islands. A Descriptive
Account of the Families, Genera, and Species indigenous to Great Britain and Ireland, with Notes as to Localities, Habitats, &c. By the Rev. CANON FOWLER, M.A., F.L.S. With two Structural Plates and Wood Engravings, complete in 5 Vols., £4. Large Paper Illustrated Edition, with 180 Coloured Plates, containing 2300 figures, £14.

A Catalogue of the British Coleoptera. By
D. SHARPE, M.A., F.R.S., and W. W. FOWLER, M.A. 1s. 6d.

The Butterflies of Europe; Illustrated and De-
scribed. By HENRY CHARLES LANG, M.D., F.L.S. Complete in Two Vols., super-royal 8vo, with 82 Coloured Plates, containing upwards of 900 Figures, cloth, £3 18s.

** THE SYSTEMATIC LIST OF EUROPEAN BUTTERFLIES *from the above work separately, price 1s.; or printed on one side of the paper only for Labels, 1s. 6d.*

British Insects. A Familiar Description of the

Form, Structure, Habits, and Transformations of Insects. By E. F. STAVELEY, Author of "British Spiders." Crown 8vo, with 16 Coloured Plates and numerous Wood Engravings, 12s.

British Beetles ; an Introduction to the Study

of our indigenous COLEOPTERA. By E. C. RYE. 2nd Edition, revised by Rev. CANON FOWLER. Crown 8vo, 16 Coloured Steel Plates, and 11 Wood Engravings, 9s.

British Bees; an Introduction to the Study of the

Natural History and Economy of the Bees indigenous to the British Isles. By W. E. SHUCKARD. Crown 8vo, 16 Coloured Plates, and Woodcuts of Dissections, 9s.

British Butterflies and Moths; an Introduction to

the Study of our Native LEPIDOPTERA. By H. T. STAINTON. 2nd Edition. Crown 8vo, 16 Coloured Plates, and Wood Engravings, 9s.

British Spiders ; an Introduction to the Study of

the ARANEIDÆ found in Great Britain and Ireland. By E. F. STAVELEY. Crown 8vo, 16 Coloured Plates, and 44 Wood Engravings, 9s.

Curtis's British Entomology. Illustrations and

Descriptions of the Genera of Insects found in Great Britain and Ireland, containing Coloured Figures, from Nature, of the most rare and beautiful Species, and in many instances, upon the plants on which they are found. Eight Vols., Royal 8vo, 770 Coloured Plates, £24.

Or in Separate Monographs.

Orders.	Plates.	£	s.	d.	Orders.	Plates.	£	s.	d.
APHANIPTERA	2	0	2	0	HYMENOPTERA	125	6	5	0
COLEOPTERA	256	12	16	0	LEPIDOPTERA	193	9	13	0
DERMAPTERA	1	0	1	0	NEUROPTERA	13	0	13	0
DICTYOPTERA	1	0	1	0	OMALOPTERA	6	0	6	0
DIPTERA	103	5	3	0	ORTHOPTERA	5	0	5	0
HEMIPTERA	32	1	12	0	STREPSIPTERA	3	0	3	0
HOMOPTERA	21	1	1	0	TRICHOPTERA	9	0	9	0

"Curtis's Entomology," which Cuvier pronounced to have "reached the ultimatum of perfection," is still the standard work on the Genera of British Insects. The Figures executed by the author himself, with wonderful minuteness and accuracy, have never been surpassed, even if equalled. The price at which the work was originally published was £43 16s.

Harvesting Ants and Trap-door Spiders; Notes
and Observations on their Habits and Dwellings. By J. T.
MOGGRIDGE, F.L.S.   Illustrated.   With SUPPLEMENT, 17s.
The Supplement separately, cloth, 7s. 6d.

Insecta Britannica; Diptera. Vol. III. By
FRANCIS WALKER, F.L.S.   8vo, with 10 Plates, 25s.

The Structure and Life History of the Cockroach
(Periplaneta Orientalis).   An Introduction to the Study of
Insects.   By L. C. MIALL, Professor of Biology in the Yorkshire
College, Leeds, and ALFRED DENNY, Lecturer on Biology in the
Firth College, Sheffield.   Demy 8vo, 125 Woodcuts, 7s. 6d.

ZOOLOGY.

Foreign Finches in Captivity. By ARTHUR G.
BUTLER, Ph.D., F.L.S., F.Z.S.   Complete in One Vol. Royal 4to,
with 60 Coloured Plates, cloth, gilt tops, £4 14s. 6d.

The Physiology of the Invertebrata. By A. B.
GRIFFITHS, Ph.D., F.R.S.E.   Demy 8vo, 81 cuts, 15s.

British Zoophytes; an Introduction to the Hy-
droida, Actinozoa, and Polyzoa found in Great Britain, Ireland,
and the Channel Islands.   By ARTHUR S. PENNINGTON, F.L.S.
Crown 8vo, 24 Plates, 9s.

Handbook of the Vertebrate Fauna of Yorkshire;
being a Catalogue of British Mammals, Birds, Reptiles, Amphi-
bians, and Fishes, found in the County.   By WILLIAM EAGLE
CLARKE and WILLIAM DENISON ROEBUCK.   8vo, 8s. 6d.

Handbook of the Freshwater Fishes of India;
giving the Characteristic Peculiarities of all the Species known,
and intended as a guide to Students and District Officers.   By
Capt. R. BEAVAN, F.R.G.S.   Demy 8vo, 12 Plates, 10s. 6d.

The Zoology of the Voyage of H.M.S. *Samarang*,
under the command of Captain Sir Edward Belcher, C.B., during
the Years 1843-46.   By Professor OWEN, Dr. J. E. GRAY, Sir J.
RICHARDSON, A. ADAMS, L. REEVE, and A. WHITE.   Edited by
ARTHUR ADAMS, F.L.S.   Royal 4to, 55 Plates, mostly coloured,
£3 10s.

ANTIQUARIAN.

A Manual of British Archæology. By CHARLES
BOUTELL, M.A. Second Edition. 20 Coloured Plates, 9s.

Sacred Archæology; a Popular Dictionary of
Ecclesiastical Art and Institutions from Primitive to Modern
Times. By MACKENZIE E. C. WALCOTT, B.D. Oxon., F.S.A.,
Precentor and Prebendary of Chichester Cathedral. 8vo, 15s.

MISCELLANEOUS.

Respiratory Proteids. Researches in Biological
Chemistry. By A. B. GRIFFITHS, Ph.D., F.R.S.E. 6s.

Collections and Recollections of Natural History
and Sport in the Life of a Country Vicar. By the Rev. G. C.
GREEN. With Woodcuts from Sketches by the Author. 6s.

West Yorkshire; an Account of its Geology, Physical
Geography, Climatology, and Botany. By J. W. DAVIS, F.L.S.,
and F. ARNOLD LEES, F.L.S. Second Edition, 8vo, 21 Plates,
many Coloured, and 2 large Maps, 21s.

Natal; a History and Description of the Colony,
including its Natural Features, Productions, Industrial Condition
and Prospects. By HENRY BROOKS, for many years a resident.
Edited by Dr. R. J. MANN, F.R.A.S., F.R.G.S., late Superin-
tendent of Education in the Colony. Demy 8vo, with Maps,
Coloured Plates, and Photographic Views, 18s.

St. Helena. A Physical, Historical, and Topo-
graphical Description of the Island, including its Geology, Fauna,
Flora, and Meteorology. By J. C. MELLISS, A.I.C.E., F.G.S.,
F.L.S. In one large Vol., super-royal 8vo, with 56 Plates and
Maps, many coloured, 21s.

The Geologist. A Magazine of Geology, Palæont-
ology, and Mineralogy. Edited by S. J. MACKIE, F.G.S., F.S.A.
Vols. V. and VI., each with numerous Wood Engravings, 15s.
Vol. VII., 7s. 6d.

The Artificial Production of Fish. By PISCARIUS.
Third Edition. 1s.

Everybody's Weather-Guide. The use of Meteoro-
logical Instruments clearly explained, with directions for securing
at any time a probable Prognostic of the Weather.   By A. STEIN-
METZ, Esq., Author of "Sunshine and Showers," &c.   1s.

Meteors, Aerolites, and Falling Stars.   By Dr. T.
L. PHIPSON, F.C.S.   Crown 8vo, 25 Woodcuts and Lithographic
Frontispiece, 6s.

The Young Collector's Handy Book of Recreative
Science.   By the Rev. H. P. DUNSTER, M.A.   Cuts, 3s.

The Royal Academy Album; a Series of Photo-
graphs from Works of Art in the Exhibition of the Royal Academy
of Arts, 1875.   Atlas 4to, with 32 fine Photographs, cloth,
gilt edges, £6 6s.; half-morocco, £7 7s.
    The same for 1876, with 48 beautiful Photo-prints, cloth, £6 6s.;
half-morocco, £7 7s.   Small Edit. Royal 4to, cloth, gilt edges, 63s.

Manual of Chemical Analysis, Qualitative and
Quantitative; for the use of Students.   By Dr. HENRY M. NOAD,
F.R.S.   New Edition.   Crown 8vo, 109 Wood Engravings, 16s.
Or, separately, Part I., "QUALITATIVE," New Edition, new
Notation, 6s.; Part II., "QUANTITATIVE," 10s. 6d.

SERIALS.

The Botanical Magazine.   Figures and Descriptions of New and Rare Plants.   By D. PRAIN, C.I.E., LL.D.,
F.R.S.   Monthly, with 6 Coloured Plates, 3s. 6d.   Annual subscription, post free, 42s. in advance.
    Re-issue of the Third Series, in Monthly Vols., 42s. each; to Sub-
scribers for the entire Series, 36s. each.

The Potamogetons of the British Isles.   By
ALFRED FRYER, A.L.S.   Royal 4to.   4 Coloured Plates, 7s.

Monograph of the Genus Teracolus.   By E. M.
BOWDLER SHARPE.   Demy 4to.   4 Coloured Plates, 7s. 6d.

Lepidoptera Indica.   By Col. C. SWINHOE.   In
Parts, with Coloured Plates, 15s. each.

THE VICTORIA LIBRARY.

A New Series of Standard and Popular Works,
in handy pocket volumes, cloth, yellow edges, 1s. each.

Vol. I., BRITISH ORATORY, containing Six famous Speeches,
viz.: Grattan on Irish Independence, Pitt on Union, Peel on
Corn Laws, Bright on Reform, Jones on Democracy, Gladstone
on Oaths.

Vol. II. ENGLISH DRAMAS : The Birth of Merlin, and Thomas
Lord Cromwell.

Vol. III. ON THE STUDY AND USE OF HISTORY : By Lord
Bolingbroke.

Vol. IV. ENGLISH DRAMAS : By Congreve. "The Way of the
World," and "The Mourning Bride."

Vol. V. A TALE OF A TUB : By Dean Swift. With notes and
translations.

Vol. VI. SPENSER'S FAIRY QUEEN : A selection of the most
beautiful passages in modernized orthography, with analyses of
each book. Notes and explanations of archaic words.

Vol. VII. LIFE OF WILLIAM PITT : By T. Evan Jacob, M.A.
Vol. VIII. ELIZABETHAN SONGS AND SONNETS.

PLATES.

Floral Plates, from the Floral Magazine. Beauti-
fully Coloured, for Screens, Scrap-books, Studies in Flower-painting,
&c. 6d. and 1s. each. Lists of over 1000 varieties, One Stamp.

Botanical Plates, from the Botanical Magazine.
Beautifully-coloured Figures of new and rare Plants. 6d. and 1s.
each. Lists of over 3000, Three Stamps.

FORTHCOMING WORKS.

Corals and Atolls. By T. WOOD JONES, B.Sc.,
F.L.S. In the press.

Flora of Tropical Africa. Vol. VI. In the press.

Flora Capensis. Vol. V. In the press.

London :
LOVELL REEVE & CO., LIMITED,
PUBLISHERS TO THE HOME, COLONIAL, AND INDIAN GOVERNMENTS,
6, HENRIETTA STREET, COVENT GARDEN.

LONDON: PRINTED BY WILLIAM CLOWES AND SONS, LIMITED,
DUKE STREET, STAMFORD STREET, S.E.

www.ingramcontent.com/pod-product-compliance
Lightning Source LLC
Chambersburg PA
CBHW021516210326
41599CB00012B/1282